T0153865

Vocabulary related to the technological
and scientific development of the 20th century
in the Tajik language on the basis
of the World Wide Web resources –
selected examples

Tomasz Gacek

Vocabulary related to the technological and scientific development of the 20th century in the Tajik language on the basis of the World Wide Web resources – selected examples

KSIĘGARNIA AKADEMICKA

Kraków

Reviewer:
prof. dr hab. Kinga Paraskiewicz

Proofreading:
Maciej Jarczyk

Cover design:
Paweł Sepielak

Publication financed by The Institute of Oriental Studies,
Jagiellonian University in Kraków

ISBN 978-83-7638-531-0

KSIĘGARNIA AKADEMICKA
ul. św. Anny 6, 31-008 Kraków
tel./faks: 12 431 27 43, 12 421 13 87
e-mail: akademicka@akademicka.pl

Księgarnia internetowa:
www.akademicka.pl

Table of Contents

1. Vocabulary related to the technological and scientific development of the 20th century in the Tajik language on the basis of the World Wide Web resources – selected examples 9
   1.1 Introduction 9
   1.2 The analysed vocabulary 10
   1.3 Sources and methodology 12
   1.4 The Tajik language and its relation to Darī and Fārsi 14
   1.5 Other languages 18
   1.6 Objectives 21
   1.7 The current state of research 23
   1.8 Some terminological remarks 24
       1.8.1 borrowings and loanwords 24
       1.8.2 coordinate compounds 24
       1.8.3 subordinate compounds 25
       1.8.4 foreign word 25
       1.8.5 idiom 25
       1.8.6 izofat chain 25
       1.8.7 izofat phrase 25
       1.8.8 semantic borrowing or borrowed meaning 26
       1.8.9 coordinative phrases 26
       1.8.10 stable izofat phrases 26
       1.8.11 scientific terminology and vocabulary 27
       1.8.12 vehicular language 27
   1.9 Writing systems, transcription and transliteration 27
   1.10 Figures 29

2 Vocabulary Analysis 31
   2.1 AIDS 31
   2.2 Airplane 33
   2.3 Allergy 38
   2.4 Antibiotic 39
   2.5 Artificial satellite 41

2.6 Atomic bomb 43
2.7 Automated Teller Machine (ATM) 46
2.8 Automatic wind-shield wipers.. 47
2.9 Black box.. 48
2.10 Blood type 49
2.11 Bluetooth technology 50
2.12 CD-ROM.. 50
2.13 cell phone (mobile phone)... 51
2.14 Cellophane.. 52
2.15 Cluster bomb 53
2.16 Computer... 54
2.17 Computer file... 55
2.18 Computer memory.. 56
2.19 Computer mouse 57
2.20 Digital still camera 58
2.21 Electric guitar 59
2.22 Electric refrigerator 59
2.23 Electric shaver... 62
2.24 Electric vacuum cleaner 63
2.25 Electronic calculator.. 64
2.26 Electron microscope 65
2.27 E-mail... 65
2.28 Flamethrower 67
2.29 Floppy disk 68
2.30 Geiger-Müller counter.. 68
2.31 Hair dryer (electric, hand-held ~) 69
2.32 Hearing aid 70
2.33 Helicopter... 70
2.34 Hormone.. 72
2.35 Insulin... 73
2.36 Integrated circuit 74
2.37 Isotope 74
2.38 Laser... 75
2.39 Lie detector... 76
2.40 Microprocessor... 77
2.41 Microwave oven.. 78
2.42 Monitor (= computer display) 79
2.43 Mp3 player 80
2.44 Neutron... 80
2.45 Nuclear power station.. 81
2.46 Nylon... 83
2.47 Prion 84
2.48 Short Message System (SMS).. 85
2.49 Superconductor... 85
2.50 Tank 86
2.51 Tape recorder.. 89
2.52 Television 90

2.53 Traffic lights 91

2.54 Transistor.. 93

2.55 Transparent Adhesive Tape 93

2.56 Vacuum tube... 94

2.57 Vitamin 95

2.58 Webcam 96

2.59 World Wide Web 96

3 Conclusions... 99

4 Alphabetical Index 103

5 Bibliography 107

5.1 Form of bibliographical citations 107

5.2 Linguistic works... 109

5.3 Sources on the history of technology and science............................. 113

5.4 Linguistic materials 117

Schemes 133

DOI: 10.12797/9788376385310.01

1. Vocabulary related to the technological and scientific development of the 20th century in the Tajik language on the basis of the World Wide Web resources – selected examples

1.1 Introduction

This work is dedicated to the study of a particular part of the vocabulary of the Tajik language that is the primarily nominal lexis related to the scientific and technological development of the 20th century. The criteria applied in the process of selecting forms chosen to be analysed are given in the section The analysed vocabulary below.

The book is focused on the Tajik (Точикӣ /tožiki/) language (henceforth: TJ). Nevertheless, the vocabulary of two other varieties of Persian (or – should the reader prefer it – two other closely related languages[1]): Fārsi (FA, i.e. the Persian as it is spoken in Iran) and Darī (DA, spoken in Afghanistan) is considered – though on a lesser scale – for comparison, too. Apart from the three closely related idioms mentioned above, lexical material of other languages has been included, to some extent, as well. These languages and the rationale for the choice of them is given in the section Other languages (p. 18).

Having defined briefly the object of the research, let us pay attention to the sources and the methodology applied. As it has been indicated in the title of this work, the most important type of sources used are electronic on-line publications. Both the sources and methodology of this work are discussed in detail in the Section Sources and methodology (p. 12).

What now remains to clarify is the aim of this work. The basic objective is an analysis of representative exemplary forms belonging to the selected part of the modern Tajik vocabulary. Etymological, word-formational, semantic and other factors will be taken into consideration in this process. The aim of the book is presented in details in the Section Objectives (p. 21).

---

[1] The problem of the relation between the three idioms is discussed on p. 14ff.

1.2 The analysed vocabulary

This work is dedicated to the study of a particular sphere of the vocabulary of the Tajik language. This sphere may be generally defined as the lexis related to the scientific and technological progress of the 20th century. To be more precise and practical, it focuses on the nominal lexical items referring either to inventions or discoveries made between the years 1901 and 2000.

Thus, to be included in the present study, a vocabulary item must be a name of a discovery or an invention. The date of the discovery or invention must be contained between the year 1901 and 2000. Where a family of words related to a discovery or invention exists (e.g. 'lie detection' as a technique and 'lie detector' as a device), only one of them is included. And last but not least only items used in everyday (i.e. non-specialist) language are analysed. Thus, the term 'laser' is accepted, while 'maser' (not used by non-specialists) is not[2].

Probably the last of the mentioned criteria deserves further explanation. Notions taken from professional jargons are not included in this study, as these jargons are in fact specific taxonomical systems with their own rules, different than those of a general language. In particular, the scientific terminology is not created in an arbitrary way. This is so, because it is created with a particular purpose, namely the purpose of classifying the phenomena specific for a given discipline. As a result, its highly organized nature makes it much more predictable than any natural language is [Baker 1998: 252]. Of course, both scientific and technological terminology sensu stricto may influence natural languages and this way it may come into our sphere of interest anyway. This phenomenon is of great importance, as the vocabulary entering the general language in this way inherits certain features of specialist terminology. One of these features is the fact that scientific terminology spreads around the world "through a small number of vehicular languages, e.g. English, French and Japanese" [Baker 1998: 252-253]. To sum up, wherever we speak of technological or scientific vocabulary in this work, words related to inventions and discoveries but belonging to the general language are meant, unless explicitly stated otherwise.

Of course, in spite of application of the above-mentioned rules, the selection of a form may be open to dispute. However, it has to be stressed that history of science and technology is certainly only a secondary discipline, as far as this book is concerned, the primary being linguistics. Moreover, as the author did his best to avoid errors in this secondary field, he does hope, potential isolated mistakes will not affect significantly the general results of the research conducted.

---

[2]   Practically, the dilemma whether to include a given form was solved by answering questions like: 1. "Is the word used in popular (not scientific or technical) acts of communication (publications, conversations, etc.)?", 2. "Is the word used in many different disciplines of sciences or technology?", 3. "Is it likely that the word is introduced in the compulsory levels of education?"

Let us present now the rationale behind the selection of this particular sphere of lexis, which is analysed in this work. First of all, it is particularly suitable for studying the newest trends in the vocabulary development, as it is relatively easy to provide a *non-ante* date for most of the words analysed[3]. The ability to estimate the earliest possible date of the forms discussed gives us the possibility to focus on the particular period (the 20th century in this case). It is a widespread opinion that most new words in modern languages are associated with technology or science[4]. Moreover, the development of this particular sub-set of vocabulary is particularly dynamic, as new phenomena to be named appear all the time in large quantities.

Another reason for choosing this particular semantic field for the research is the fact that words associated with technology are significantly less culture-dependent in comparison to some other subsets of vocabulary. Even if we find specific examples like **online confession** (clearly rooted in the Christian context) or **electronic Adhan** (obviously of Islamic origin), still number of such phenomena is relatively small in comparison to other spheres of vocabulary. This makes the vocabulary associated with technology and/or science better comparable, and thus best suited for a comparative research.

An interesting paradox is that although the vocabulary in question in not culture-dependent, still possessing the up-to-date set of technological (and scientific) lexica is in a sense important for the culture of the speakers of a language, as it is a paragon of the idiom's vitality and ability to describe the modern world. As Misra and others put it in their book on the roots and consequence of deprivation, "*... it is true that in a technological culture the language with technological vocabulary is in an advantageous position...*" [Misra et al. 1995: 124][5]. For some languages which have not developed this field of lexica naturally, there is an on-going effort to provide such vocabulary artificially, e.g. Breton [Hale & Payton 2000: 60]. Moreover, lack of technological or scientific vocabulary in a language may be one of the factors leading to abandoning a language in favour of a better equipped rival [Hindley 1990: 217]. This way we find that studying technological vocabulary of a language may tell us a lot about its present state and prospects for the future. In this context one should remember that Tajik indeed has still a powerful rival. More than 10 years after the declaration of independence the process of introducing Tajik into state institutions was still not very advanced. The same was true about some scientific institutions [Johnson 2006: 171].

One more remark should be made referring to the analysed corpus of vocabulary. No strict differentiation between scientific and technological vocabulary is made in this work, as the author's research was not focused on this subject. However, we

---

[3] Instances of words pre-dating an invention or discovery (like the already mentioned **robot** or **helicopter**) are exceptions.

[4] This is true esp. for EN [Burgmeier et al. 1991: 4].

[5] It is worth noting that they express their uncertainty, whether technological culture is universally a good thing.

have to remember that certain differences between the two sets of vocabulary do exist. E.g., the technological terminology is, in general, more heterogeneous and easier influences the common language [See Baker 1998: 252].

1.3 Sources and methodology

As the title of the present work suggests, on-line electronic texts are among its most important sources. Their significance derives from the relative simplicity and affordability of publishing a text on the Internet together with the speed of this process. All that makes them particularly important in the research focused on the fast changing sphere of vocabulary. We also have to keep in mind that Internet-based materials become an increasingly important source for works dedicated to linguistic (especially lexical and word-formational) research [see e.g. Mühleisen 2010: 1].

Nevertheless, other types of publications have been extensively used, most of them belonging to the following categories:

☞ dictionaries and other lexicographical sources,

☞ scientific publications providing data on the lexis of TJ and other languages taken into consideration,

☞ scientific, popular and other publications dedicated to the scientific and technological development of the 20th century.

☞ other publications containing the vocabulary important for the present work (e.g. schoolbooks).

Let us focus for a moment at the lexicographical sources, as these are crucial for the type of research conducted by the author. Among these, the following publications may be named:

☞ Eršov N.N., et al. 1942. Luḡat-i Harbi-i Rusi-Toǧiki [Eršov 1942];

☞ Bertel's E.È., et al. (eds.) 1954. Tadžiksko-Russkiy slovar' [Bertel's 1954];

☞ Ja. Kalontarov. 1955. Kratkiy tadžiksko-russkiy slovar'. Moskva. Izdatel'stvo Inostrannẏx i nacional'nẏx slovarey.

☞ Šukurov M.Š., et al. (eds.) 1969. Tolkovẏy slovar' tadžikskogo yazẏka w 2-x tt. Moskva. Izdatel'stvo Sovetskaya Ènciklopediya.6

☞ Osimi M., Arzumanov S.D. 1985. Luḡat-i Rusi-Toǧiki [Osimi & Arzumanov 1985];

☞ Moukhtor Ch. et al. 2003. Dictionnaire tadjik-français [Moukhtor et al. 2003];

☞ Saymiddinov D., et al. (eds.) 2006. Farhang-i toǧiki ba rusi [Saymiddinov et al. 2006];

☞ Ja. Kalontarov. 2008. Farhang-i nav-i toǧiki-rusi (Kalontarov 2008);

☞ Nazarzoda S., et al. 2008. Farhang-i tafsiri-i zabon-i toǧiki [Nazarzoda et al. 2008]

---

6 This dictionary is not used in the present work, as it focuses on the pre-20th century TJ and – as such – was not particularly useful.

☞ F. Sanginov. 2010. Luḡat-i anglisi ba rusi-vu toӡiki va rusi ba anglisi-vu toӡiki. Grammatika (Sanginov 2010)

Other sources include mass media publications and broadcasts, and a limited number of literary works (in isolated cases even poetry, which is normally not a typical source for technological vocabulary).

The starting point for the research was establishing a list of lexical meanings, for which TJ equivalents have been sought. This is a method applied earlier by Haspelmath and Tadmor in the Loanword Typology Project [Haspelmath & Tadmor 2009: 5]. As the mentioned authors stress, a meanings list is not simply a word list in some language (EN in particular), even if it is usually presented in this way [Haspelmath & Tadmor 2009: 5]. It is rather a list of ideas. Items from this list are used as section titles of this work.

Of course, there is no guarantee that for every item in the meanings list it will be possible to find a single-word lexical equivalent in TJ (or some other idiom). In fact, such an equivalent will be often a word group [Cf. Haspelmath & Tadmor 2009: 11]. Thus, studying the terminology in question both word-formation and syntax are applied.

Structural differences are not the only ones and certainly not the most important ones. It is a well known linguistic phenomenon that what we see as lexical equivalents in any two languages should not be expected to have perfectly the same semantic content. Probably the best known example is that of colour names in various languages. In other words, lexical equivalents are seldom perfect on the semantic level, too. We have to keep this in mind when discussing the material of this study.

For all the meanings included, the *non-ante* date is presented, which is normally a year of discovery or invention, as it seems reasonable to assume that the word related to it did not exist previously or, at least, it was not commonly known[7]. In other words, it could be assumed that only after that moment its expansion to numerous languages started.

All the forms are analysed etymologically (diachronic analysis) and word-formationally/syntactically (synchronic analysis). In the case of loanwords, an important element of the analysis is indication of the donor language, or – to be precise – donor languages, as we should make difference between the immediate and original source. This is a complicated issue, especially in the case of numerous internationalisms appearing in this sphere of vocabulary. In some cases indicating one single immediate source may be easy (e.g. the TJ form **гормон** 'hormone', where the initial consonant indicates RU as the vehicular language), however, in other instances it may be difficult or even impossible [cf. Hapelsmath & Tadmor 2009: 16], e.g. TJ **хуликуптар**.

Another important feature in the case of loanwords is the level of phonological and morphological integration of a form with the system of the recipient language

---

[7] There are examples of inventions that were ignored at first and had to be 're-invented' even decades later (see e.g. the section on automated teller machine, p. 46 ff.

[Cf. Hapelsmath & Tadmor 2009: 16]. It is not only a piece of information inter-esting per se, but also an important indication whether we are dealing with a real borrowing, or rather a foreign word used in a text. In fact, distinguishing loanwords from code-switching is one of the important problems of the research on lexical bor-rowing [Treffers-Daller 2010: 17].

The analysed TJ vocabulary items are compared to equivalents in other idioms – first of all FA and DA, then other languages of the region, as well as possible do-nor languages (see The Tajik language and its relation to Darī and Fārsi below. And Other languages, p. 18). Taking this into consideration, the present work belongs – at least partially – to the sphere of lexical typology, as it deals – among other things – with certain patterns (tendencies to borrow words, typical source languages etc.) observable in the given field of vocabulary [Koptjevskaja-Tamm 2008: 4-6]. In par-ticular, the present study touches the problem of borrowability in the given field of vocabulary in a number of idioms [Koptjevskaja-Tamm 2008: 6].

A number of auxiliary techniques have been used in the study, e.g. to estimate the popularity of forms in question. An Internet search or web-corpus investigation [see e.g. Mühleisen 2010: 1] is one of them. To assure uniformity and comparability of the results, one search engine have been used (Google). In some cases, additional criteria were formulated (as Google does not allow, for now, choosing Tajik as the language of the search, limitation of the results to domains registered in Tajikistan – .tj was applied).

1.4 The Tajik language and its relation to Darī and Fārsi

It is a disheartening phenomenon indeed that what remains in the centre of any lin-guistic study, i.e. the language itself, escapes a good and unquestionable definition. As a consequence, there is no indisputable classification of the languages of the world, even if the vast majority of scholars agree on most of the problems [Majew-icz 1989: 9]. It is even worse when it comes to decide, whether a particular idiom is a language indeed or just a dialect of another one. This question, in the author's opinion, may not be answered by linguistics alone. Strict linguistic rules to solve the 'language or dialect' problem could be (and have been) developed (basing on quan-tity of isoglosses, mutual intelligibility or any other empirically observed factors), but they are impractical, as they contradict our commonly accepted scientific clas-sification, popular knowledge, and self-identification of speakers etc.[8] Any criteria

---

[8]   Even within the Iranian world we find examples of this phenomenon, e.g. Kurdish-speakers speaking Sorani and Kurmanji dialect (?) generally express their belief in existence of one Kurdish language, even if the mutual-intelligibility criterion would rather make us classify these ethnolects as separate languages. Similarly, speakers of Arabic dialects of the opposite extremities of the Arab-speaking zone (e.g. from Maghrib and Yemen) may fail to understand each other. On the other hand, Maltese is classified as a separate language.

that would try to comply with these factors, on the other hand, would surpass the limits of linguistics and subjective interpretation of certain aspects would be inevitable.

One of the cases, where there is a lot of hesitation between the terms 'a language' and 'a dialect' is the problem of the relationship between Tajik, Darī and Fārsi, and esp. between TJ on one side and FA & DA on the other. There is no consensus about the status of TJ in the scholarly publications. E.g. Kerimova believes the differentiation between TJ and FA started as early as the 16th century and it was concluded in the 20th with forming a new literary standard of TJ [Kerimova 1997: 96ff.]. Kerimova is surely right in noticing the process of differentiation and it is unquestionable that this tendency gained a lot of impetus in the 20th century, when what was to become Tajikistan got effectively isolated from Afghanistan and Iran, while Russian and Uzbek influences were reinforced. And according to Comrie, political separation may increase the level of linguistic differentiation between idioms [Comrie 1981: 9]. The same author mentions explicitly TJ and FA as being rather different from one another, especially in abstract and technical vocabulary because of the intensive inflow of borrowings from RU in TJ [Ibid.][9].

However, were the changes deep enough to change the dialect into a language? Perry acknowledges the fact that the differentiation of TJ and FA started at least in the 16th century [Perry 2005: 1], but – at the same time – he uses the name 'Tajik Persian' for the language he describes. Moreover, he presents an idea of a continuum of Persian dialects with TJ and FA as its opposite extremities and the dialects of Afghanistan in between [Perry 2005: 1]. One should not forget at this place that in the 20th century the differentiation of Tajik was not only a natural process (in the unnatural, isolated environment of the Soviet Central Asia), but much more a result of a planned language policy [Perry 2005: 1-3][10]. Publications (scientific ones included) of the Soviet period must be read with awareness of the fact that they are not only descriptions of TJ, but also instruments of shaping the opinion about the language, tools of language policy [cf. Perry 2005: 3]. In other words, they may be rather tools used to change the reality, rather than to describe it. How politics, ideology and linguistics intermingled in the Soviet period is well presented by Grenoble in the passage on the history of the Marxist doctrine [Grenoble 2003: 55ff.].

An interesting statement on the status of TJ was made by Bacon: '… Tajik, a language so closely akin to Persian that Tajiks claim it is Persian.' [Bacon 1980: 27]. In fact, TJ native-speakers of today and the present language policy makers of the independent Tajikistan are not uniform in their opinions about the status of TJ, as it is proven by some hesitation in the republic's language legislation. According to the

---

[9]   Comrie says that FA, differently from TJ, borrows mostly from AR [Comrie 1981: 9]. This seems to be imprecise, esp. nowadays, as the influence of FR and EN on FA is – e.g. in the case of the technical vocabulary – much stronger.

[10] "In the space of a mere two generations, Tajik has been one of the most consciously, intensively, and rapidly 'planned' languages ever – both a the stage of russianization (late 1920s to 1950s) and again during re-persianization (late 1990s onward)" [Perry 2005: 2].

language law of 1989, the state language of Tajikistan was "Tajik (Farsi)". This "Farsi" in parentheses, however, was dropped in 1999 [Fierman & Garibova 2010: 440].

Taking all this into consideration, the author of the present work feels not in a position to solve the 'language or dialect' dilemma in reference to the status of Tajik. Thus, another way out will be taken. Neutral terms as 'idiom' and 'ethnolect' will be used referring to any of the three: TJ, FA & DA.

Of course, where the distance between two languages grows, the problem of classifying them as dialects or separate languages disappears. As it has been already said, in the past centuries the distance between TJ and FA (and DA) was growing constantly with great the acceleration in the last century. One of the objectives of this book is to estimate the tendencies in the vocabulary of the three idioms: Do we notice sings of further differentiation or is there any tendency to preserve (regain) unity?

Having discussed the 'language or dialect' dilemma, we may try to define the Tajik idiom. The term itself, as it is often so with linguonyms, is an ambiguous (but nevertheless useful) notion. Using the name 'Tajik' one may refer either to: (1) the local descendant of the Classical Persian, (2) to 'a modern written variety of internation Persian', (3) or to a group of dialects spoken in the region [Perry 2005: 2]. Also (4) the so called Modern Literary Tajik (MLT) may be meant that is an artificially created standard of literary TJ, based extensively on the northern dialects [Perry 2005: 3].[11]

Unfortunately, the terminology used in reference to these particular idioms is not universally accepted and consistent. What is mentioned in point (1) is often referred to as 'literary language'. Of course, this must not be confused with the MLT. In fact, the idiom of the point (1) is of little interest taking into consideration the subject of the present work, as it is 'virtually unaffected by the trappings of modernity' [Bashiri 1994: 120].

One has to admit that the creation of MLT must have been a hard work. A more or less consistent system was to be designed out of a number of dialects. Moreover, as the new standard idiom was meant to be used as an official one (beside Russian), further obligations were imposed on the creators and executors of the Soviet language policy in Tajikistan. Thomason mentions a number of things that have to be done, when a newly official language is adapted, including codifying a standard, writing/printing official documents and teaching materials, promoting the new standard, creating and popularizing words for situations not referred to in the given idiom before, etc. [Thomason 2001: 38-30].

To see, how complicated the relations between the four varieties of TJ may be, esp. when political agenda is present, let us have a look at a passage in the work

---

[11] In some cases the word 'Tajik' is also (erroneously) used to designate an eastern Iranian ethnolects properly called Sarikoli and Wakhi, which are spoken in western China [Perry 2009: 1041]. As Iranian language, they are related to Persian, however, much more distantly than DA and TJ (not closer than, for instance, Pashto). This use of the linguonym Tajik will be excluded from our further considerations.

of Kerimova and Rastorgueva of 1964, dedicated to the problem of the use of the verbal prefix **би-** in TJ. The authors first declare that it is used mostly in poetry and – sometimes – in artistic prose. This refers clearly to the type of TJ we have defined in point (1). Then they say that it is a characteristic feature of southern and part of the central dialects, while it is practically absent from the northern ones. This, of course, should be qualified under the point (3). The authors also say that it is not used in the spoken language of intelligentsia ("(…) *в разговорном языке интеллигенции она отсутствует.*" [Rastorguyeva & Kerimova 1964: 27]. This 'spoken language of intelligentsia' can be identified with the MLT (4). Moreover, the use of the word 'intelligentsia' here seems to imply that using this form of language ennobles the speaker, while speaking southern or central dialects definitely does not[12].

In the present work, 'Tajik' is understood mostly in the sense of the point (4) or (2) and in most cases the choice between the two will be obvious, as the sources of discussed forms will be given explicitly. The meaning of the point (3) is present to certain extent, as results of the research conducted among the TJ-speakers and the study of some other sources (like unofficial, personal electronic publications) may show some traces of dialects.

On the other hand, it would not be correct to perceive the varieties of TJ as separate idioms, isolated from each other. There is certainly a lot of mutual influence between them. Moreover, it is important to note that the native speakers refer to all these types of TJ as to one language, and not a group of languages.

As a consequence, we have to look at TJ as a complex phenomenon rather than a single idiom. This makes it extremely different to make general statements on what the differences between TJ and FA (also TJ and DA) are. Observations that are valid for MLT (4) are not necessarily so in the case of the local descendant of Classical Persian (1), not to mention 'modern internation Persian' (2). The dialects (3) make the picture even more complicated, as they are a separate universe, highly differentiated internally.

Nevertheless, we may present certain features specific at least for some of the types of TJ. As far as the phonology is concerned, the differentiation of vowel length present in Classical Persian was rejected in both TJ and FA, however, the results are different. In TJ a new opposition (unknown in FA) of stable and unstable vowels appeared [Perry 2009: 1041-1042]. Even if the differences between FA and TJ in the sphere of phonology are not very great, still they are significant enough to make the TJ vowel system practically identical with that of Uzbek [Perry 2009: 1041]. Of course, in the case of northern dialects of TJ, the influence of Uzbek is much more intensive and it reaches far beyond the sphere of phonology [Schlyter 2006: 815], see e.g. the use of the question enclitic **-ми** /-mi/. Certainly, one of the most important differences between TJ and FA is to be found in the morphology of verbs. Even though the basic inventory of tenses is similar in both idioms [Perry 2009: 1042], TJ uses a lot of forms unknown in FA, just to mention continuous tenses with the

---

[12] See Perry for a more balanced description of the use of the prefix **би-** [Perry 2005: 234-235].

auxiliary verb **истодан** /istodan/, numerous non-witnessed forms, etc. [Perry 2009: 1042]. Nevertheless, one may find a great deal of TJ texts abstaining from using any verbal forms not known in FA (See e.g. the official Tajik translation of the Universal Declaration of Human Rights). These texts may be described as written in 'modern internation Persian' (2).

As far as the vocabulary is concerned, the scale of differentiation is particularly high in the case of MLT. It is probably the variant of TJ, where Russian loanwords and calques are particularly numerous (see section Other languages, p. 22). These differences are of high importance considering the subject of this book. Thus, TJ forms will be often confronted against FA and DA ones, which will hopefully help reconstruct their way into the language.

1.5 Other languages

Other languages, whose lexical material is taken into consideration in the present work may be divided into two classes:
☞ languages spoken in the same regions as TJ, FA and DA are, esp. those with large number of speakers (particularly when they are bilingual, speaking TJ/FA/DA as well), those possessing official status and/or a literary standard.
☞ Languages being the main sources of loanwords for TJ, FA and DA.

For understandable reasons, the way the vocabulary of the two categories mentioned above is used in the present work in different ways.

In the case of TJ, languages belonging to the first class are mainly Russian (also class two) and Uzbek[13]. In the case of the latter one has to remember that the northern dialects of TJ, themselves highly influenced by UZ, were crucial in creating the literary standard of modern TJ [Ido 2005: 5; Grenoble 2003: 152]. Another Turkic language exerting some influence on TJ is Kyrgyz (KY) [Grenoble 2003: 152].

In the case of DA, **самолёт** Pashto has to be taken as a representative of the first category, while as far as FA is concerned, we should take into consideration e.g. Azeri Turkish (AZ).

As far as the second class is concerned, the difference in the sources and inventory of borrowings in TJ and FA is a well known phenomenon [Megerdoomian 2009: 296]. Generally speaking, English and – to a lesser extant – German and French play an important role here, at least in the case of FA and DA while TJ is supposed to borrow a lot of lexical items from RU. One has to remember that from the moment of tsarist Russia's expansion in the Central Asia until the end of the Soviet empire, RU was used as an official state language in Tajikistan. Even if national languages (TJ among them) were officially recognized in the USSR, none of them

---

[13] Esp. northern dialects of TJ are influenced by UZ. This phenomenon is stimulated by the widespread TJ-UZ bilingualism in the areas where these dialects are spoken [Perry 2005: 3].

enjoyed prestige and the universality of RU[14]. As early as in 1960's there were about 2,500 RU loanwords in TJ including those used to express ideas associated with military techniques, medicine, agriculture etc. [Bashiri 1994: 118]. In some cases, RU words were introduced even in the case where a native equivalent existed, e.g. **лампа** 'a lamp' for **чароғ** [Bashiri 1994: 118; Bacon 1980: 197].

As far as the earlier research in the field is concerned, Bacon gives some statistics regarding the presence of RU loanwords in TJ in the period around the middle of the last century. Basing on the various dictionaries published then she states that it is the scientific vocabulary where the rate of RU loanwords was particularly high. However, as Bacon notices, the proportions varied significantly between different disciplines. In the case of mathematics, the rate of the RU borrowings was relatively low and there over 20% of them and 24% more words of mixed TJ and RU origin. In the case of biology and agriculture, the proportions were in favour of the TJ forms, as well. However, in the case of physics, native words formed below 28.5% of the vocabulary. In the case of the chemical terminology, TJ forms constituted around 17% of the whole corpus, and mixed forms slightly below 20% [Bacon 1980: 196][15].

At this point we have to make an important remark that the results obtained for some particular sphere of vocabulary do not necessarily reflect precisely the situation of the lexical corpus as a whole. A good example of this phenomenon is the comparison of the amount of loanwords from German and French in various spheres of the Japanese vocabulary: scientific and technological vocabulary is borrowed mostly from German, while French is dominant in other parts of vocabulary (like arts and literature). Moreover, analysing the entire vocabulary, French is slightly ahead of German [Dohlus 2010: 150-152].

To this we should add the factors resulting from the policies of the Soviet period, effective isolation from the related idioms (DA & FA) being the most important. This certainly strengthened the role of RU as the most important source of loanwords in TJ in the 20th century[16]. Moreover, the social relations, ideology, politics, economy

---

[14]   In fact, all the three idioms in question came under the influence of RU (as a consequence of the political influence of Russia) at certain point: in the case of FA this influence (starting in the 19th century) was relatively weak and nowadays only about 100 RU loanwords are attested in this language [see Bashiri 1994: 110]. in the case of DA, this influence became observable after the Soviet invasion in 1979. As far as the TJ is concerned, the impact of RU was stronger and more persistent.

[15]   As Bacon herself notices, these were the results of the study of dictionaries 'designed for specialists' [Bacon 1980: 196], so they are not comparable, in a straight-away manner to the material studied in this work (let us remind that we focus on the words associated with science and technology, but only those known to or used by the common public).

[16]   Not questioning this assessment as such, we have to remember about two important problems: 1. the question of persistence of the RU influence on TJ (esp. now, when some factors inducing it are no longer in action), 2. the question to what extent sources of the Soviet era are reliable in their description of this particular phenomenon, rather than being paragons of some politically inspired wishful thinking.

of the Soviet empire were significantly different in many aspects from the rest of the world, and this different reality required a lot of new words, the most natural source of them being, again, RU. Examples like **колхоз** [Saymiddinov et al. 2006: 279], **интернатсионализм** [Saymiddinov et al. 2006: 247] are a good illustration of the problem[17].

RU became a natural source in the case of the lexis related to scientific and technological development. Comrie refers to this phenomenon a number of times. At first he states that: "Russian had already assimilated much of the western technology and culture, and so did not have the problem, faced by many other languages of the USSR, of first coming to terms with these phenomena (...)" [Comrie 1981: 31]. He also indicates vocabulary of the languages of USSR as the field where the influence of RU is particularly visible, mentioning explicitly technical vocabulary in particular [Comrie 1981: 33]. Comrie notices, too that "most of the languages of the USSR had no technological vocabulary of their own referring to elements of western technology or culture, but borrowed such words from Russian." [Comrie 1981: 35].

Moreover, promoting borrowing from RU instead of creating new lexical items basing on the existing lexical items was a part of the internal consolidation policy of the Soviet authorities. It is worth noticing that there was a tendency to retain original (i.e. RU) orthography (even if it was not consistent with the one devised for a given language) and the pronunciation as close to RU as possible for the speakers of the idiom in question [Comrie 1981: 34; FTAF 1941: 13][18]. We could mention here the word **стакан** /stakan/ 'a glass (in the sense of a container)' with the initial consonant cluster normally not possible in most variants of Persian (cf. FA استکان/estakān/, being the loanword from RU, too, however, adapted to fit the phonology of the idiom). Another example is the word **самолёт** one of the forms for 'an airplane' used in TJ, which is pronounced, at least by part of the speakers, [samal'ot], i.e. in accordance with the original RU phonetics of the word (see p. 34ff.).

Another important factor strengthening the role of the most important source of loanwords was the policy of the central Soviet authorities to promote bilingualism, involving always RU – as the language of the whole country – and a local idiom.

The next phenomenon we should pay attention to is the fact that nearly all of the written languages of USSR used Cyrillic alphabet, even if for some of them other

---

[17] It is worth noticing that the influence of RU on other languages of USSR had no parallel in their impact on RU. It is hardly possible to find words borrowed from other languages into RU, apart from those referring to local phenomena of non-RU speaking Soviet republics [See Comrie 1981: 31].

[18] Although this rule was formulated expresis verbis [**Филиали Точикии Академияи Фанхои СССР** 1941: 13], still it was not applied entirely consistently. Some letters of the Cyrillic alphabet used to write down RU and redundant in TJ were retained in accordance with the rule in question, e.g. **ц** (as in **цирк** or **цемент**) [**Филиали...** 1941: 15-16]. Other, however, were not used in TJ, e.g. **щ** was to be substituted with the sequence **шч** (as in **мешчанин**) [**Филиали...** 1941: 17].

writing systems were employed at first (including TJ). This facilitated e.g. borrowing scientific terms from RU basing on written texts.

One has to remember that the phenomenon of borrowing lexical items – which is crucial for our analysis – is not the only way the languages interact and some types of this interaction may obscure the results of a study like the present one. We should mention here code-switching, i.e. mixing the material of two (or more) languages (codes) in a single conversation [Thomason 2001: 132], and using unassimilated foreign words (see section 1.8.4 on p. 25). Generally speaking, conversational material (especially in mixed language) is not analysed in the present work so the problem of code-switching is not an important one. On the other hand we shall see a lot of examples of foreign words, which may be interpreted as a specific example of code--switching (mixing is limited to introduction of isolated foreign words into a text).

1.6 Objectives

As it has been already said in the section The analysed vocabulary, p. 10, the main objective of this book is to analyse examples of forms belonging to the particular sphere of the vocabulary of the Tajik language. This encompasses both diachronic and synchronic aspect, as both the etymology and the word-formational structure of the forms in question will be discussed. Semantic and syntactical problems will be studied as well.

One of the questions we are going to answer is that about the proportion between terms based on native vocabulary and loanwords. All the idioms in question take part in numerous inter-language contacts and – in such situation – influence of one language on another is inevitable. Now, borrowing of words is "the most common specific type of influence" of this sort [Thomason 2001: 10].

Nowadays, one expects most languages of the world to borrow words associated with the scientific or technological development from a limited set of privileged idioms, especially EN. This intuition is true in most cases, however, not always. Thomason gives examples of North American Indian languages (Montana Salish, Nez Percé, Sahaptian) that – despite being under heavy influence of EN, as most of their speakers are bilingual – tend to create new lexical items basing on native vocabulary, rather than borrowing them from EN [Thomason 2001: 11][19]. Thus, the percentage of loanwords in the analysed lexical corpus cannot be estimated a priori and a language change (including borrowing of a lexeme) is not a predictable phenomenon. The fact that the contact between some languages makes it possible, does not make it obligatory [see Thomason 2001: 77].

---

[19] By chance, Thomason provides evidence that this is true specifically about the words associated with the technological development cf. Montana Salish **p'ip'úsyšn** 'an automobile' (lit. 'wrinkled feet') [Thomason 2001: 11] or Nez Percé **cewcew'in'es** 'telephone' (lit. 'thing for whispering') [Ibid.].

Still, borrowing a word is one of the most common contact-induced language changes. It may happen even in case of casual, not very intensive contacts. Other kinds of contact-induced language changes (e.g. borrowing of syntactical or morphological structures) may occur only if the contact is intense, especially in bilingual societies [Thomason 2001: 78, 80]. Thus, we may expect TJ to borrow words from UZ, RU and from EN. However, although we notice numerous examples of structural borrowings from UZ (like the question-indicating enclitic -ми /-mi/ [Perry 2005: 294]) and even some from RU (like the sequences of adjectives joined without the *izofat* morpheme, e.g. **кадрхои ташкилӣ-партиявӣ** /kadrho-i **taškili-partiyavi**/ 'party-organizational staff' [Perry 2005: 488]). However, to the author's best knowledge, we find no structural borrowings from EN.

As far as the borrowings are concerned, we are going to answer a number of questions in the present work:

What are the most important sources of the borrowings (in the sense of original or primary source-languages)?

☞ What are the most typical vehicular (or intermediary) languages?

☞ What is the typical number of vehicular languages in the chain of transmission of a singular form?

☞ Is it possible to indicate typical routes (single original source languages and a repeating sequences of vehicular languages)?

☞ Are there any observable tendencies for the change of the typical sources of borrowings?

As the basic objectives of the present work are fulfilled, the results obtained will be used to draw further conclusions. In particular the author would like to estimate the impact of particular languages (Fārsi, Dari, Russian, Uzbek, English etc.) on the vocabulary of modern Tajik. This may help to indicate tendencies in the development of lexis of the three closely related idioms. That such tendencies may be observed, shows the work of Bacon in reference to TJ in the first half of the 20th century, who noticed increase in the number of RU loanwords in the idiom in question [Bacon 1980: 197].

Of course, we have to be aware of the fact that the tendencies observed regarding lexical items related to science and technology do not have to appear in other spheres of vocabulary. Especially proportions of loanwords vs. native forms may be different. This is not only supported by the common intuitive belief that the probability of borrowing a word is related to the semantic sphere it belongs to [Haspelmath & Tadmor 2009: 1], but it may be proved by statistical analysis [Haspelmath & Tadmor 2009: 7]. Still, restricting the material of research only to the words used by non--specialist speakers should make the results more representative.

Analysing tendencies is particularly interesting in the case of TJ. Even in the beginning of the 1900's it was noted that in some of the new publications authors purposely abstained from using RU loanwords [Bashiri 1994: 133]. Having said this, let us recall the fact that analysis of the tendencies in vocabulary development has to be conducted along two different paths. First of all, we have to focus on the nature of the new forms. However, the situation of the forms previously incorporated into the lexical system deserves some attention, as well. One has to remember that especially the new words are often – after some time – replaced by their equivalents of different nature: loanwords by native words or other loanwords[20]; native words by loanwords[21] [Cf. Haspelmath & Tadmor 2009: 13]. In fact, every new word exerts some effect on the pre-existent lexicon: it may be described either as replacement, insertion or coexistence[22].

1.7 The current state of research

Even if the number of publications dedicated to the study of TJ is incomparably lower to those focusing at FA, still there is considerable number of works dealing with this idiom. Let us present a selection of these publications.

Some general descriptions of the language are available, starting with A. Semenov, *Kratkiy grammatičeskiy očerk tadžikskogo yazỳka (...)*, Taškent 1927. As far as more recent grammatical sketches of TJ are concerned, one may mention the chapter on TJ by A.A. Kerimova in the multi-volume work on the languages of the former Soviet Union (*Yazyki Narodov SSSR*, t. 1, Moskva, 1966). G. Lazard in his 'Caractères distinctifs de la langue tadjik' (published in *Bulletin de la Société de Linguistique de Paris* 52/1, 1956, pp. 117-86) focuses on the specific features of TJ.

The development of TJ as a language (dialect?) independent from Persian is presented in a number of publications, just to mention J.R. Perry, 'From Persian to Tajik to Persian: culture, politics and law reshape a Central Asian language' (*in Non-Slavic Languages 8. Linguistic Studies in the Non-Slavic Languages of the Commonwealth of Independent States and the Baltic Republics*, ed. H. I. Aronson, Chicago, 1996, pp. 279-305); L. Rzehak, *Vom Persischen zum Tadschikischen. Sprachliches Handeln und Sprachplanung in Transoxanien zwischen Tradition, Moderne und Sowjetmacht (1900-1956)*, Wiesbaden, 2001).

Among the recent descriptions of the grammatical system of TJ works by J.R. Perry (*A Tajik Persian Reference Grammar*, Leiden, 2005), H. Karimov and others (*Grammatika-i zabon-i adabi-i hozira-i tožik*, Dušambe 1985) may be mentioned.

As far as the TJ lexicographical sources are concerned, various dictionaries published during the last one hundred years are of particular importance. The list of such publications is provided in the chapter on the sources of this work (see p. 12ff.).

Works dedicated to the Tajik lexicon in particular exist, too. Some of them deal specifically with the problem of loanwords from other languages in TJ, e.g. I. Bashiri,

---

[20] See the TJ ← RU ← FR аэроплан practically replaced by a number of equivalents: самолёт (← RU), хавопаймо (← FA), тайёра (← AR).

[21] See лампа vs. чароғ – p. 23

[22] Haspelmath & Tadmor made such an observation referring to loanwords [Haspelmath & Tadmor 2009: 16], however, it is equally valid for new native forms, too.

'Russian Loanwords in Persian and Tajiki Languages' (in *Persian Studies in North America: Studies in Honor of Mohammad Ali Jazayery*, ed. M. Marashi, Bethesda, Md., 1994, pp. 109-39) and G. Doerfer, *Türkische Lehnwörter im Tadschikischen* (Wiesbaden, 1967). There are also some publications dealing with the TJ lexica of the post-Soviet period, e.g. *Sud'by rusizmov v tadžikskom Yazyke postsovetskogo perioda* by È. Sobirov (Sobirov 2007).

On the other hand, the author of this book has been unable to find a publication presenting results of a research parallel to his own.

1.8 Some terminological remarks

1.8.1 borrowings and loanwords

Borrowing, in the widest possible sense, is understood as "the incorporation of features of one language into another (...)" [Treffers-Daller 2010: 17]. A specific type of *borrowing* is a lexical one. If not explicitly declared otherwise, this particular type of *borrowing* will be normally meant wherever the term is used. Moreover, the form *loanword* will be used as a synonym to *lexical borrowing. Loanword* is a lexeme transferred from a donor language into a receiver language, and used in the latter as a word [Haspelmath & Tadmor 2009: 13]. One has, however, to remember that in the literature the form *loanword* is sometimes used in the broad sense as an umbrella term for both *foreign word* (q.v.) and *borrowing* (i.e. the form lexicalised into the destination language) [Bussmann 1998: 702].

1.8.2 coordinate compounds

Coordinate compounds are the ones with the coordinative relationship between their constituents, i.e. "both concepts are predicated equally of the same referent" [Olsen 2004: 17; 18]. Examples of TJ coordinate compounds are **ширқахва** 'coffee with milk' (/šir/ 'milk' + /qahva/ 'coffee') and **пуропур** 'overfull' (/pur/ 'full' + linking morpheme /o/ + /pur/). Forms like **обу хаво** with the enclitic conjunction /-u/ shall not be classified as coordinate compounds (contrary to Perry [Perry 2005: 441ff.]), as their elements seem to retain original stresses, i.e. they are not single words, and so they are not true compounds, even if they share some of their features (e.g. the ability to form plurals). They are rather specific word groups and they will be referred to in this work as *stable coordinate [word] groups (phrases).*

1.8.3 subordinate compounds

The type of compound, in which one element is subordinate to (dependant on) the other one. This is a vast category comprising such sub-types as descriptive, possessive and determining compounds. The subordinate element will be called 'the modifier' while the other one will be labelled as 'the head'. In the case of DA/FA/TJ it cannot be said a priori which element is the head and which one is the modifier.

1.8.4 foreign word

A form absorbed by a language from another, which – contrary to a *borrowing* or *loanword* (q.v.) – has not been phonetically, graphemically or grammatically assimilated to fit the destination language system. The difference between a *foreign word* and a *borrowing* may often be unclear and disputable. [Bussman 1998: 420]. In the present work, all the forms that are not assimilated in one of the mentioned aspects (phonetically, graphemically or grammatically) will be classified as foreign words. As the number of *foreign words* defined in this way is much lower than that of *loanwords,* only the previous ones will be explicitly marked.

1.8.5 idiom

The word *idiom* is used in the present work mostly to refer to languages and/or dialects abstaining from declaring what their status in fact is.

1.8.6 izofat chain

See *izofat phrase*.

1.8.7 izofat phrase

Izofat phrase is a typical construction used in TJ (just like in FA and DA) to link a word (in most cases a noun) with its modifier (typically an adjective or a noun) [Perry 2005: 71-72]. An unstressed enclitic particle **-и** /-i/ is added to the head word [Ibid.]. Apart from joining a word with is modifier, *izofat* is also extensively used in DA/FA/TJ syntax. Subsequent modifiers may be added to already existing *izofat phrases* forming *izofat chains*. When a modifier in the previous *izofat* phrase is a noun, the subsequent element may refer to both the head and to the modifier of the previous phrase. In certain cases one may argue that one *izofat* phrase is the head (or a modifier) of another one, e.g. in the FA phrase قرون وسطی اروپای غربی 'The Middle Ages of (in) Western Europe' it seems reasonable to classify the phrase *qorun-e vostā* as the head of the larger izofat construction.

As a result, *izofat chains* may be quite complicated in their structure and meaning (see examples like **аломати норасоии масунияти бадан** or **бемории норасоии муҳассали масуният**).

Izofat chains are often multi-dimensional structures contracted as to make it possible to convey them through the communication channel lacking this feature of multi-dimensionality. For example the Persian izofat chain[23] **otāq-e kuček-e zir-e širvāni-ye Ali** (an example provided by Moinzadeh 2006: 48) may be presented as follows[24]:

We have also to say that an izofat phrase is a word group, not a word, with its elements retaining their original stress. However, in some cases such phrases become stable or even petrified and even obtain a meaning which is not easily deductible from its elements (e.g. **чаъбаи сиёҳ** 'a black box' – q.v. or **шарҳи ҳол** 'biography'). Thus, such phrases permanently enrich the lexicon of the language (see *stable izofat phrases*).

1.8.8 semantic borrowing or borrowed meaning

Borrowing meaning from one language to another without actually transferring the word [Haspelmath & Tadmor 2009: 13]. In practice, there usually is a pre-existing word, which changes its meaning (or acquires some additional meaning) under the influence of a foreign word or idea [see Bussmann 1998: 138].

1.8.9 coordinative phrases

These are petrified word groups joined by the enclitic conjunction /-u/, cf. **coordinate compounds**.

1.8.10 stable izofat phrases

Some izofat phrases become petrified or even lexicalised (e.g. **тухми мурғ** /tuxm-i murǧ/ "an egg"), a phenomenon on the border between syntax and word-formation. As they retain original primary stresses they are certainly not single words (compounds), even if they sometimes act like these (e.g. they may serve as the basis for further affixal derivation). C.f. **subordinate compounds**.

---

[23] There are no important structural differences between the izofat construction in FA and TJ.
[24] See the section on graphs conventions.

1.8.11 scientific terminology and vocabulary

By scientific and/or technological terminology specialist taxonomic systems (or jargons) will be meant, while scientific and/or technological vocabulary will be used to denote part of the lexical corpus of the general language related to science and/or technology.

1.8.12 vehicular language

A vehicular language is the one that is the immediate source of a foreign word in some language, but it is not the original source of it.

1.9 Writing systems, transcription and transliteration

The transcription used for TJ forms follows the system used by J. R. Perry [Perry 2005]. One significant difference is that the voiced palatal affricate is rendered by /ǯ/. The palatal resonant – in its turn – is written as /y/[25]. This alternations help to retain coherence in romanization of various Cyrillic orthographies (TJ and RU in the first place – see below).

Non-distinctive features are not shown – as a rule – in transcription, e.g. the glide [-y-] in an intervocalic position is not marked[26] and regular phonetic changes, like devoicing of /b/ before voiceless consonant (e.g. **хубтар** /xubtar/ 'better' actually pronounced [xuptar] [Perry 2005: 23]). Whenever a more precise, phonetic rendering of pronunciation is needed, it is written in square brackets, e.g. [bombaʸi atomi], [samal'ot], while the phonemic transcription is encircled by slash brackets, e.g. /tuxm-i murḡ/).

	Front	Central	Back
Close	i		u
Mid	e	ů	o
Open	a		

Table 1: Vowel System of Tajik

---

[25] The RU letter ы (used to render the central vowel /i/) is transliterated as <ẏ>.

[26] A glide appears for instance when a vowel-starting morpheme is added to a vowel-ending word, e.g. **бомбаи атомй** transcribed /bomba-i atomi/, actually pronounced [bombaʸi atomi].

	Labial	Alveolar	Palatal	Velar	Uvular	Glottal
Stops	p, b	t, d		k, g	q	?
Affricates			č, ǯ			
Fricatives	f, v[27]	s, z	š, ž	x, ḡ	h	
Nasals	m	n				
Resonants		l	y			
		r				

Table 2: Consonant system of Tajik

Transcription of FA and DA, where used, differs in details from the system pre-sented above, e.g.: In FA, [q] and [ḡ] are just allophones of one phoneme /q/. The open back rounded vowel /ɒ/ (historically corresponding to the TJ /o/) is transcribed in the present work, in accordance to the wide-spread practice, as /ā/. In the case of DA, vowel length (still preserved there) is marked.

Forms taken from other works and transcribed with alphabets different from Lat-in are normally romanized, e.g. in the case of Lebedev (1961), where Pashto words are rendered using the Cyrillic alphabet. Original transcription will be used only in the case where the interpretation of the form is dubious.

The Cyrillic script of the forms discussed is not entirely identical with the system used at the moment in the Republic of Tajikistan. The TJ orthography at the mo-ment does not use some letters that were peculiar to RU loanwords (ц, щ, ь, ы), however, the author of the present work has retained them, wherever they appear in analysed forms (esp. those pre-dating 1998). On the other hand, the letter ў used in modern TJ publications issued in the Republic of Uzbekistan is substituted with ӯ, its counterpart in the modern TJ orthography. However, the letter ў will be preserved in UZ language forms written in the Cyrillic script. Forms belonging to some other languages of the former Soviet Union are given in the Cyrillic alphabet as well, like UZ, AZ, KY etc. The orthography applied follows the original sources from which these forms are taken so it may be – in some cases – inconsistent. Moreover, in the case of some of the languages written with Cyrillic, Latin alphabet was/is used as well, so forms in both writing systems coexist.

The Perso-Arabic script is used mostly in its variant specific to FA, DA and TJ (today only marginally). Some sources give Perso-Arabic writing for forms that en-tered TJ vocabulary after effective abandonment of this writing system by Tajiks at the beginning of the 20th century. These forms are cited as well, as they are particu-larly interesting, because they are comparable to the FA/DA counterparts and they reveal in some aspects more than the Cyrillic alphabet (esp. the – nowadays rather

---

[27] This phoneme has a positional bilabial allophone in TJ [Perry 2005: 24-25]. However, as it has no phonemic value, this shall not be reflected in the transcription.

hypothetical – length of vowels). Forms in the Arabic script are also given for some other languages, like Pashto.

Description of methods of transcription and transliteration used in bibliographical entries and references is provided in the chapter 5.

1.10 Figures

Various types of graphs appear in this book. The first is constituted by graphs showing the sequence of word-formational and syntactic processes leading to appearance of particular forms. Let us call them '**sequence graphs**'. They are an element of synchronic analysis, the results of which may sometimes differ significantly from the outcome of etymological (diachronic) analysis (e.g. certain forms taken from Iranian Persian are classified as borrowings etymologically, nevertheless they are word-formationally analysable in TJ).

Etymology graphs show the origin of a particular form from a diachronic perspective. (diachronic analysis) and word-formational graphs (synchronic research).

Relation graphs show a relation between a modifier (or a complement) and a head in a phrase (an izofat phrase in particular).

Some general rules will be followed in both types of graphs, namely, independent words will be drawn as ellipses, bound morphemes as boxes and other word-groups as polygons. Other forms – if appearing – may be drawn as diamonds. In Etymology graphs, forms of foreign origin are placed in triangles. Additional information on the nature of relation between the elements of the graphs may be shown on the edges connecting the objects.

DOI: 10.12797/9788376385310.02

2 Vocabulary Analysis

2.1 AIDS

AIDS is an acronym for Acquired Immunodeficiency Syndrome. This disease was for the first time conclusively identified in the US in 1981 [Encyclopaedia Britannica vol. I, 67; Messadié 1995: 230].

A number of related descriptive forms are used in TJ to designate AIDS:

The first one to discuss is **бемории пайдошудаи норасоии масуният** [Jumhuriyat acc. 2014-09-07]. It may be literally translated as 'disease of acquired deficiency of immunity'. It is an izofat chain well attested in online resources (comparing to the rest of the studied forms) [Google search: keyword="бемории пайдошудаи норасоии масуният", date=2014-09-07]. Another form is **аломати норасоии масунияти бадан** [Maǵlis-i Oli 2004-12-09] /alomat-i norasoi-i masuniyat-i badan/ (lit. 'syndrome of deficiency of the immunity of the organism'). Its presence in online resources is limited to around 200 results [Google search: keyword="аломати норасоии масунияти бадан", date=2014-09-07]. This is not a precise translation of neither EN nor RU (**синдром приобретённого имунного дефицита**) term, as it lacks the notion of 'being acquired'. Structurally, it is a multi-level set of *izofat* phrases (i.e. an *izofat chain*), where the relation structure is quite simple (every subsequent word is a modifier of the previous one). an acronym based on this form, **АНМБ**, is attested [Huseynova 2010; Google search: keywords=АНМБ+AIDS[28], date=2014-09-07].

Another TJ form for **AIDS** is **бемории норасоии масунияти одам** [Karim [2011]-05-17] /bemori-i norasoi-i masuniyat-i odam/. Its attestation in online resources is rather poor [Google search: keyword="бемории норасоии масунияти одам", date=2014-09-07]. It is structurally identical to the form discussed above. It differs only in using the word **бемор**ӣ 'disease' instead of **аломат** 'syndrome'. We have to keep in mind that in the professional medical terminology 'syndrome' and 'disease' are not synonyms. However, in the popular (as opposed to scientific) language their meaning is very close if not identical, especially in a phrase like the one discussed. Also the word-formational/syntactic structure of both forms is very similar.

---

[28]   The keyword АНМБ alone produces unrelated results in RU.

норасоии масунияти одам [Karim [2011]-05-17] /norasoi-i masuniyat-i odam/ [Barotov & Gulxoja 2010-12-01; Ašurova 2011-05-11] is another variant, this time lacking both 'syndrome' and 'disease'. **бемории норасоии мухассали масуният** [Firûz & Šarif acc. 2011-06-08] has neither **бадан** /badan/ nor **одам** /odam/ as one of its elements, but – on the other hand – it contains the form **мухассал** /muhassal/, which renders the notion of 'being acquired'. It is poorly attested online [Google search: keyword= "бемории норасоии мухассали масуният", date=2014-09-07]. It seems to be a good translation of the RU **синдром приобретённого имунного дефицита**, if we agree to accept TJ **беморй** as a valid rendering of the RU **синдром**. Just like in the previous cases, we have got an izofat chain here. However, the relation structure of this phrase is a bit different (see *Fig. 3*).

The next form is **аломати мухассали масунияти одам** [Nigori & Munavvar acc. 2011-06-08]. Here 'being acquired' is clearly linked to **аломат** 'syndrome', like in RU. However, on the other hand, it lacks the form **норасой** /norasoi/, which is – in fact – quite essential. It is not very popular in online resources [Google search: "бемории норасоии мухассали масуният", date=2014-09-07].

Some more related variants like **бемории норасоии бадан** /bemori-i norasoi-i badan/ [Barotov & Gulxoja 2010-12-01] and **норасоии масуният** /norasoi-i masuniyat/ [Barotov & Gulxoja 2010-12-01] have been attested.

Apart from that, the RU acronym **СПИД** [Ustoyev 2008: 141; Ašurova 2011-05-11; Barotov & Gulxoja 2010-12-01] pronounced /spid/ [Mažalla-i Bomdodi 2010-07-21; Barotov & Gulxoja 2010-12-01] is used in TJ and – in fact – it is very popular. It happens to be used together with the full name in TJ (i.e. **аломати норасоии масунияти бадан**) [Mažlis-i Oli 2004-12-09]. It is popularly used and understood, so that phrase-books suggest to use it instead of other forms [Rudelson 1998: 186]. It does not seem to be well assimilated, so it will be classified as a foreign word.

The EN acronym is borrowed into TJ, too. It may be written in the Cyrillic script as **ЭЙДЗ** [BBC Persian 2010-12-01b; Radyo-i Ozodi 2006-06-01] and it is pronounced /eydz/ [Barotov & Gulxoja 2010-12-01]. Interestingly, it also happens (even if much more rarely) to be written **ЕЙДЗ** [BBC Persian 2009-11-10], which could possibly show some discrepancy in the use of the inherited RU yotated letters. Finally, the original EN acronym in the Latin script, i.e. **AIDS**, is sometimes used, too [Radyo-i Ozodi 2008-11-12]. All these derivatives of the original EN acronyms will be classified as foreign words in TJ, as they are all not assimilated phonetically (normally we do not find the consonant cluster [-dz] in this language) or both phonetically and graphemically.

The transcribed EN acronym is also used e.g. in AR: الأيْدْز [Ba'albaki 1999: 35]. In UZ we find the acronym **ОИТС,** which is based on the native expression **орттирилган иммунитет танкислиги синдроми** [Šodmonov acc. 2011-06-12], itself a calque of RU or EN form. However, let us pay attention to an interesting fact – the UZ form is phonetically quite close to the EN **AIDS.** In AZ the acronym **QIÇS** is used (based on: **Qazanılmış Immun Çatışmazlığı Sindromu**) [Cahan oğlu

Əliyev acc. 2011-09-21], however, the RU acronym **SPİD** (not the dotted capital I) is used in AZ, too (Ibid.).

In FA other forms are used: the full name is نشانگان نقص ایمنی اکتسابی/nešānegān-e naqs-e imani-ye ektesābi/ [MEPI acc. 2010-12-26] or سندرم نقص ایمنی اکتسابی [Google search: keyword= "سندرم نقص ایمنی اکتسابی", date=2014-12-23] and the acronym borrowed from EN is attested: ایدز /eydz/ [Haghshenas et al. 2002: 28; MEPI acc. 2010-12-26]. Apart from that, an adaptation of the FR acronym **SIDA**, i.e. سیدا is attested, too [Google search: keyword= "سیدا", date=2014-12-23]. In DA, the form ایدز is used [Killid Group 1389-09-10 HŠ; ARXA acc. 2010-12-25] /aydz, eydz/ [Awde et al. 2002: 29; 71].

In PŠ the EN acronym is used, too in the form: ایډز [BBC Pashto 2010-12-01; Google search: keyword= ایډز, date=2010-12-25][29].

2.2 Airplane

Although the dream to fly is one of the archetypical motives of human culture and various attempts have been made before the beginning of the 20th century (with intensiveness peaking in the 19th century), it is commonly accepted to attribute construction of the first airplane sensu stricto[30] to Wright brothers, their successful attempt being dated 1903.

The form **аэроплан** /aeroplan/ appears in the earliest studied sources [Eršov et al. 1942: 15; FTAF 1941: 18; Bertel's et al. 1954: 37], those written thirty years ago [Osimi & Arzumanov 1985: 42] and the most recent analysed lexicographical works [Moukhtor et al. 2003: 14]. AR script variants of this form are ائراپلن [Nazarzoda et al. 2008: 1,106] and آیرپلان [Rzehak 2001: 71]. On the other hand, some of the new TJ dictionaries (e.g. [Saymiddinov et al. 2006) do not mention the form **аэроплан** at all. In fact, it seems to be quite rare in modern use: An Internet search reveals less than 100 results, about five of them in TJ [Google search: keyword=**аэроплан**; domain=.tj; date=2010-11-09]. Indeed, Nazarzoda marks this form as an old one [Nazarzoda et al. 2008: 1,106].

Plural form of **аэроплан**, i.e. **аэропланхо** /aeroplan-ho/ is attested, even if scarcely (An Internet search [Google search: keyword=**аэропланхо**, date=2011--01-01] reveals a single result [Hamad 2009-06-29]). Also, the derived form **аэроплансозй** [FTAF 1941: 31] /aeroplan-soz-i/ 'airplane construction, production' is used (see *Fig. 4*).

Another attested derivative is **аэропланчй** [Bertel's et al. 1954: 37] /aeroplan-či/, in which the agentive suffix of Turkish origin **-чй** /-či/ [Perry 2005: 422] is added to the noun in question.

---

[29]   Retroflexed stops /ṭ/ and /ḍ/ regularly render the EN alveolar /t/ and /d/ [Penzl 1961: 47].

[30]   I.e. heavier than air, self-powered, controlled aircraft.

A related form, آئروپلان, is known in FA [Omid 1373: 1211], too, however it is not very popular[31]. For DA, no traces of the form in question have been found neither in lexicographical works, nor in the Internet resources [Google search: keyword=آئروپلان, domain=.af, date=2010-11-13].

The word **аэроплан** is an internationalism. It is attested in many languages (RU included) and its primary source is the FR **aéroplan** [Tokarski et al. 1980: 8; Černyx 1999: 1,61], where it was coined basing on the GR elements **ἀήρ** 'air' and **πλάνος** 'wandering' [Woodhouse 1972: 961], Thus, 'wandering in the air'. In FR it has been used since the middle of the 19th century and in RU since 1880s[32]. It seems most probable that the word was transferred to TJ from RU (cf. full orthographical correspondence).

As far as the languages of Central Asia are concerned, it certainly may be found in UZ [Koščanov et al. 1983: 45], beside other loanwords like **самолёт/samolyot, tayyora** (see below) and forms of Turkic origin: **uchoq, uchqich** [Balci et al. 2004: 298] (cf. TR **uçak**, a neologism based on the verb **uçmak** 'to fly' [Alkım et al. 1996:1194]). The form **аэроплан** is also attested in AZ (beside **тәйярә**) [Alizade et al. 1944: 9]. We also find the same internationalism in TK: **аэроплан** [Hamzaev 1962: 61] and in (Tabrizi) AZ as /aeroplan/, /ayriplan/ [Householder & Lotfi 1965: 239].

The form **самолёт** /samolyot/ is borrowed from RU and, moreover, it is one of a very limited set of forms of Slavonic origin in the analysed corpus. Its word-formational structure is clear: **само** (← **сам** 'self') + **лёт** (← **летать** 'to fly'). That RU was a direct source of this form is evident. Some Tajik speakers even preserve palatalisation of the consonant /l/ (→ [l']), a phenomenon typical for RU consonant system[33] and the etymological /o/ in the second syllable is changed into [a] (another RU phonetic process)[34] [Mažalla-i Bomdodi 2010-07-15].

Whether RU is the original source is a separate question. Parallel forms exist in other Slavonic languages (c.f. BG **самалёт**, PL **samolot**). Nevertheless, the RU form is attested quite early even if with a different meaning ('a type of a river ferry') [Afanas'ev 1861: 127; Černyx 1999: 1,61]. In TJ, just like **аэроплан**, it appears both in older [Eršov et al. 1942: 15; FTAF 1941: 13] and newer [Osimi & Arzumanov 1985: 986] lexicographical works. The form **самолёт** is not mentioned in

---

[31]   Internet search produced about 350 results for this form [Google search: keyword=آئروپلان, language=FA, date=2010-11-13], while a parallel query for one of the rival forms هواپیما produces over 13,300,000 [Google search: keyword=هواپیما, date=2010-11-13].

[32]   Let us remind that, as a principle, forms analysed in the present work have to designate inventions or discoveries made in the 20th century. The forms themselves may be older, like in this case. **аэроплан**

[33]   Some Tajik speakers tend to retain palatalization in numerous borrowings from Russian [Gacek 2012: 356].

[34]   The so called *akanye* (RU **аканье**) resulting in the change of the original unstressed [o] into [a], typical for dominating RU dialects. *Akanye* may be noticed in pronunciation of some TJ speakers [Gacek 2012: 355].

Saymiddinov's dictionary [Saymiddinov et al. 2006]. However, it is used much more often today, especially in the spoken language[35]. Written sources, including schoolbooks, use this form as well [Boinazarov et al. 2007b: 30].

The assessment of the frequency of **самолёт** in modern online TJ resources is a bit complicated, as the orthography of the word is identical with the RU original. Nevertheless, the search of this word in coincidence with specifically TJ forms like **фурудгох**, **мусофирбар** and **нишаст** results in between 200 and 600 websites [Google search: keywords= самолёт+фурудгох, самолёт+мусофирбар, самолёт+нишаст; date=2014-09-08].

It has to be stressed that the traditional Cyrillic spelling of this word reflects RU orthography and not the actual pronunciation (neither RU nor TJ).

The regular plural form of **самолёт**, i.e. **самолётхо** /samolyot-ho/ is attested [Google search: keyword=самолётхо, date=2011-01-01]. Further derivatives based on the form in question are attested, too, e.g. **самолётсозй** [Google search: keyword=самолётсозй, date=2014-03-15] /samolyot-soz-i/ 'airplane construction, production'. As far as the word-formational structure is concerned, **самолётсозй** is identical with **аэроплансозй** (See Fig. 4: аэроплансозй – etymology above).

Another form based on **самолёт** is the subordinate compound **самолётрон** 'airplane pilot' [Bashiri 1994: 119] /samolyot-ron/ (which has a rival in another borrowing, namely **авиатор** [Bashiri 1994: 127]), where the second element (the head) is the PrSS of the verb **рондан** /rondan/[36]. A further abstract derivative is to be found, as well: **самолётронй** [Bertel's et al. 1954: 339].

The form **самолёт** is also used in some stable izofat phrases like **самолёти реактивй** 'jet plane' [Bashiri 1994: 120] /samolyot-i reaktivi/.

Another interesting form is **гидросамолёт** 'hydroplane' [Bashiri 1994: 123]. The element **гидро-** 'hydro-' (from GR via RU, as it is indicated by the change of initial /h-/ into /g-/) may be encountered in many other TJ forms of RU origin, like **гидрограф** [Bashiri 1994: 123], **гидродинамика** [Saymiddinov et al. 2006: 148] etc. However, the author of the present work has been unable to find an example of a form with the very same element attached to a native one. Thus, there is no data to put forward the hypothesis that it is a productive element in TJ. In other words, it is safer to assume, the forms containing **гидро-** were all borrowed from RU in their entity[37].

The form **самолёт** is not used in DA and FA. On the other hand, it is attested in other languages of the post-Soviet Central Asia, e.g. UZ [Salihov Salihov & Ismatul-

---

[35] Baizoyev and Hayward mark this form as one of the RU forms used in the colloquial TJ [Baizoyev & Hayward 2004: 355].

[36] Bashiri classifies the morpheme **рон-** /ron-/ as a suffix, however, in the humble opinion of the author of this work it is more satisfactory to call it a verbal stem, and the whole word a compound.

[37] Forms like **гидрографй** [Bashiri 1994: 123] are obviously secondary, the suffix has been added to the borrowed **гидрограф**.

layev Ismatullaev 1990: 28; Koščanov et al. 1983: 45][38], KY [Yudahin 1957: 752], TK [Hamzaev 1962: 61].

It is quite interesting that some other languages borrowing a word for **airplane** from RU (Bezhta, Ket, Sakha), actually borrowed both of the mentioned ones, i.e. **самолёт** and **аэроплан** [WLD 2009: Meaning 23.16: the airplane, 2011-05-07]. If only one of them was taken, it was usually the previous one (like in Archi or Kildin Saami [WLD 2009: Meaning 23.16: the airplane, 2011-05-07]).

The word **хавопаймо** /havopaymo/ is a form of Iranian origin and it obeys the FA : TJ voice rules. The AR script orthographical variant هواپیما [Nazarzoda et al. 2008: 2,476] is identical with the one used in FA [Rubinèik 1970: 2,732] and DA [Yussufi 1987: 130]. As a native Iranian form, this word is not attested in RU. As far as TJ is concerned, it appears mostly in newer publications [Saymiddinov et al. 2006: 672; J. J. Rudelson 1998: 171; Baizoyev & Hayward 2003: 364].

As far as the word-formational structure of this noun is concerned, it is a subordinate compound with the noun **хаво** 'air' and the present verbal stem **паймо-** (← **паймудан** 'to travel, to traverse') as its elements. Thus, the etymological meaning of the form is 'the one that travels in the sky'.

According to Sadeghi, the FA form هواپیما was purposely coined by a society working under the auspices of the Ministry of War around 1924 [Sadeghi 2001: 21]. It is a convincing view, especially that the word seems to be a calque of the FR **aéroplan** (see above).

The TJ form seems to have developed independently and initially it had a different meaning 'a pilot, aviator' [Bertel's et al. 1954: 495] only later to change designation most probably under the influence of the FA هواپیما [Perry 2005: 489]. Indeed, even in some modern texts we find the word **хавопаймо** with personal meaning. This is particularly clear when the form is accompanied with personal plural marker **-он**, e.g. **Рӯзи хавопаймоён** /rûz-i havopaymoyon/ 'Pilots' day'. Interestingly, at least one modern dictionary mentions this personal meaning of **хавопаймо** as the only one [Kalontarov 2008: 283].

The form **хавопаймо** in the sense of 'an airplane' is a fully-functional noun and it has a regular plural: **хавопаймохо** [Google search: keyword=хавопаймохо, date=2010-12-30].

To check the presence of this form in online sources, the same method was used as for the **самолёт**, i.e. websites were looked for, in which the form in question appears together with the following: **фурудгох**, **мусофирбар** and **нишаст**. It is not necessary in the case of **хавопаймо**, as the form is specifically TJ itself, however, using the same method guarantees the comparability of the outcome. The results were between 14600 and 31200 websites found [Google search: keywords= хавопаймо+фурудгох, хавопаймо+мусофирбар, хавопаймо+нишаст; date=2014-09-08], which is by far more than in the case of **самолёт**.

---

[38] Also the romanized variant: **samolyot** [Balci et al. 2004: 242].

Further derivatives based on **хавопаймо** exist, e.g. **хавопаймой** [Saymiddinov et al. 2006: 672] /havo-paymo-i/ (= **хавопаймо** + abstract nominal suffix **-й** /-i/) 'aviation'. The homophonic adjectival suffix is also used with this form to produce a related adjective **хавопаймой** [See e.g. Tajik Air acc. 2010-12-30]. There also exists a form parallel to **самолётсозй**, but based on **хавопаймо** as its constitutive element, i.e. **хавопаймосозй** (cf. FA هواپیماسازی [Argāni 1364:15]), see Fig. 4: аэроплансозй – etymology above). Another attested derivative of the form **хавопаймо** is the adjective **хавопаймобар** /havo-paymo-bar/ (see Fig. 5 below), used in the *izofat* phrases **нови хавопаймобар** /nov-i havopaymobar/ and **киштии хавопаймобар** /kišti-i havopaymobar/ both meaning 'aircraft carrier'[39]. The form هواپیمابر is used in FA, as well [Argāni 1364: 15] beside the form with the additional adj. suffix /-i/, i.e. کشتی هواپیمابری as attested by Alizade and others [Alizade et al. 1944: 3].

Another derived form is **хавопайморабо** [Nazarzoda et al. 2008: 2, 476; Radyo-i Ozodi 2009-04-20] /havopaymo-rabo/ 'airplane hijacker'. It is a subordinate compound comprising the word **хавопаймо** as its modifier part and the verbal present stem **рабо-** /rabo-/ (← **рабудан** /rabudan/ 'to rob') as the head. The next level of derivation is also attested, namely the abstract noun **хавопайморабой** [Nazarzoda et al. 2008: 2,476; BBC Persian 2009-07-17] /havopaymorabo-i/ 'airplane hijacking' (see Fig. 6 below). Finally, we note the form **хавопаймозадагй** [Nazarzoda et al. 2008: 2,476] /havopaymo-zadag-i/ 'airsickness'.

The form **тайёра** /tayyora/ is well attested in TJ in sources of various periods of the 20th century [Eršov et al. 1942: 15; Salihov Salihov & Ismatullayev Ismatullaev 1990: 28; Saymiddinov et al. 2006: 573; Rahimov et al. 2006: 7; Mažalla-i Bomdodi 2010-07-15]. In the dictionary by Bertel's from 1954, the author marks this form as an archaic one [Bertel's et al. 1954: 376].

The form is borrowed from AR (cf. طيارة [Danecki & Kozłowska 1996: 511]), and its AR script form accepted in TJ is طیاره [Nazarzoda et al. 2008: 2,294]. The form is known in FA [Āryānpur & Āryānpur 1375: 1418; Haghshenas et al. 2002: 21] and DA [Ostrovskiy 1987: 319; Glassman 1971: 199][40], too, even if in FA it is not the most universally used word for 'airplane'. The same word appears in PŠ (طياره /tayyārá/ Lorenz 1979: 286]) apart from the native لوتکه /alwutáka/ [Lebedev 1961: 603]. It is also attested in non-Iranian languages like UZ: **tayyora** [Balci et al. 2004: 274], (Tabrizi) AZ /tæyaaræ/, /tæyyaræ/ [Householder & Lotfi 1965: 267]. This form is particularly interesting as it is relatively rare for a word of AR origin to be used in DA/FA/TJ for a notion associated with modern scientific or technological development. According to Sadeghi, the form has been introduced into FA via TR

---

[39] **Ершов** in 1942 gave only one form for 'aircraft carrier', namely a loanword from Russian **авиаматка** /avyamatka/. This might have been an ad-hoc borrowing, as he felt the necessity to give a periphrastic definition at the relevant entry [Eršov 1942: 7]. Nowadays this form seems to be obsolete in TJ, however, it is used e.g. in AZ [Alizade et al. 1944: 3].

[40] With possible pronunciations /tayāra/ and /tyāra/ [Awde et al. 2002: 66-67].

[Sadeghi 2001: 20], however, it is not clear for the author of this work what was its way into TJ.

To analyse the frequency of **тайёра** in the online sources, the form was – just like in the case of **самолёт** and **хавопаймо** – sought for in the text where it co-exists with the words: **фурудгох**, **мусофирбар** and **нишаст**. The results are up to 1100 hits [Google search: keywords= тайёра+фурудгох, тайёра+мусофирбар, тайёра+нишаст; date=2014-09-08], which means the form is a bit more popular than **самолёт**, but still it is not an important rival for **хавопаймо**.

The word is well assimilated in TJ. It forms normal plural: **тайёрахо** [Qarḡizova 2009-07-07; Vazorat-i Naqliyot 2009-03-30] /tayyora-ho/. Apart from that, compounds and derivatives containing **тайёра** as one of their constituents are also attested, e.g. **тайёрасозӣ** /tayyora-soz-i/ 'airplane construction' [Salmonov 2009-03--16] (c.f. **хавопаймосозӣ** above).

To sum up, we see that of the four words for 'an airplane' presented, only **аэроплан** is limited in its use and is perceived as an old form (even if not obsolete). The three remaining forms seem to be acceptable and in some cases interchangeable (cf. **тайёрасозӣ**, **хавопаймосозӣ** and **самолётсозӣ**). On the other hand, some difference between them is perceived. Esp. **самолёт** is recognized as a RU form, and it is omitted even in some lexicographical works [Nazarzoda et al. 2008: vol. 2].

2.3 Allergy

In 1906 Clemens von Pirquet introduced the term allergy, while studying skin reaction to tuberculin [Pirquet 1906: 1457; Messadié 1995: 23; Jaffuel et al. 2001].

The form **хассосӣ** [Saymiddinov et al. 2006: 686] /hassosi/ is used (and had been used before acquiring the additional meaning of 'allergy') in a much more general sense of 'sensitivity'. In other words, we have got a pre-existing word here that has changed its meaning (or – to be precise – acquired additional meaning) under the influence of a foreign word or idea. Thus, it is a classical example of *borrowed meaning* [Bussmann 1998: 138] or semantic borrowing [Haspelmath & Tadmor 2009: 13]. The term **хассосӣ** itself is derived from the AR حَسَّاس /ḥassās/ 'sensitive' (← <ḤSS>) accompanied by the TJ abstract suffix **-ӣ** /-i/. Interestingly, in AR the term for allergy is based on the same lexeme, to which the native abstract suffix is added: حَسَّاسِيَّة /ḥassāsiyya/. The resulting form – just like the TJ **хассосӣ** – is also used with the meaning of 'sensitivity' etc. [Danecki & Kozłowska 1996: 289]. In FA and DA the adapted AR variant حساسيت /hassāsiyyat (hassāseyyat)/ is used [Rubinčik 1970: 1,503; Ostrovskiy 1987: 29], again both with the meaning of 'allergy' and 'sensitivity'. See also the DA adj. /hasās/ 'allergic' [Awde et al. 2002: 42].

In TJ the word **хассосият** /hassosiyat/ is used with the meaning 'allergy', too [Osimi & Arzumanov 1985: 30; Rasulov 2010-03-10]. Again this medical meaning is a new one at it has been acquired in the 20th century. The form differs from **хассосӣ** in being an AR abstractum borrowed en bloc (cf. AR حَسَّاسِيَّة above).

The words **аллергӣ** [Moukhtor et al. 2003: 6] /allergi/ and **аллергия** (الیرگیه in the AR script) [Nazarzoda et al. 2008: 1,55] /allergiya/ are both variants of the same internationalism first used in DE (**Allergie** [Pirquet 1906: 1457-1458]). It is derived from the GR ἄλλος 'other' and ἔργον 'action, work, etc.' [Tokarski et al. 1980: 20; Jaffuel et al. 2001; Groves 1834: 27 & 244]. The first of the TJ forms has been adapted by placing the abstract suffix **-ӣ** /-i/ at its end, while the latter show clearly RU as its immediate source (the ending /-(y)a/ being a marker of one of the RU nominal declensions). In the case of **аллергӣ** RU as an intermediary is very probable, too, even if the adaptation is deeper. Let us consider the fact that the form simillar to **аллергӣ** is attested in FA: آلرژی /ālerži/ [Omid 1373: 43; Haghshenas et al. 2002: 34] and in DA آلرژی /aliržī/ [Lebedev et al. 1989: 45]. Comparing them to the TJ form, we come to the conclusion that it is the stop /-g-/ in the TJ form that suggests RU as the immediate source in the case of the TJ form, while the spirantic /-ž-/ in the FA and DA variants points rather at FR.

The second of the two related forms, **аллергия**, is a borrowing with no trace of adaptation. Parallel forms are to be found in other languages of the region, cf. UZ **allergiya** [Balci et al. 2004: 12], AZ **allergiya** [Öztopçu 2000: 362], etc.

A related form is attested in PŠ الرژی /alerži/ [Lebedev et al. 1989: 45] and in TR **alerji** [Alkım et al. 1996: 48], however, again (as in the case of FA/DA), the spirant /-ž-/ indicates FR as the vehicular language for these words.

A derivative of either **аллергӣ** or (more probable) **аллергия** constructed using the adjectival suffix **-(в)ӣ** /-vi/ is attested: **аллергиявӣ** /allergiyavi/, e.g. in the phrase **беморихои аллергиявӣ** [Karimzoda 2010: 23] /bemoriho-i allergiyavi/. A compound adj. **аллергияовар** [Nisaun 2009-03-09] /allergiya-ovar/ 'causing allergy' with the present stem of the verb **овардан** /ovardan/ 'to bring' as its head is attested, too.

Interestingly, beside the form cited above, AR uses variants of the very same internationalisms, too, namely أَرَجِيَّة, أَلِيرْجِيَّة [Arslanyan & Šubov 1977: 28].

2.4 Antibiotic

Though antibiotics are released naturally into soil by bacteria and fungi, they did not come into worldwide prominence until the introduction of penicillin in 1941 [Encyclopaedia Britannica vol. 1, 449] *discovered by Flemming in 1928* [Brzeziński 1995: 358; Rembieliński & Kuźnicka 1987: 133]

The form **антибиотик** [Osimi & Arzumanov 1985: 34; Saymiddinov et al. 2006: 41; Ustoyev 2008: 102; Moukhtor et al. 2003: 9] /antibyotik/ is a typical internationalism and it was most probably borrowed into TJ as a whole. Nevertheless, we should note that the foreign morpheme **anti-** has the status of semi-prefix (or prefixoide) in TJ [Rzehak 2001: 357].

The primary source of the form in question is the FR **antibiotique**, a pseudo-classical borrowing based on the GR αντì 'opposite' + βίος 'life, existence' [Tokarski et al. 1980: 37; Groves 1834: 53 & 114] and accompanied by the FR adjectival suffix. The term was first used by Dubos in 1940, however, it was modelled upon **l'antibiote**, invented by Vuillemin in 1889 [Phan 2008: 11].

The TJ **антибиотик** is well attested in online resources [Mahkamova acc. 2014-09-10; Sang 2012-05-07; Sahimov 2008].

Similar forms are used in FA, آنتی‌بیوتیک [Omid 1373: 52; Asadullaev & Peysikov 1965: 32] /āntibyotik/ and in DA انتی‌بیوتیک pronounced /antibeyōtik/ [Ostrovskiy 1987: 30] or /antibayātik/ [Awde et al. 2002: 28]. We note that the traditional sound correspondence between TJ /o/ and FA /ā/ is not occurring here, which makes the possibility of FA serving as a vehicular language a bit less probable. Comparing the FA & DA form, we come to a conclusion that the immediate source for DA might have been EN (see the structure of the third syllable), while the FA form was modelled on some other language, most possibly FR.

In FA, apart from آنتی‌بیوتیک we also find forms like پادزی [Āryānpur 1375: 617] / pādzi/ [Mirzabekyan 1973: 36] /pād-ziv/ and پادزیست [Haghshenas et al. 2002: 49].

Forms identical to that attested in TJ are to be found in RU and other languages of the post-Soviet Central Asia (e.g. RU **антибиотик**, UZ **антибиотик** [Koščanov et al. 1983: 34]). Thus, one feels entitled to suppose that RU was the immediate source here.

The plural form **антибиотикхо** /antibyotik-ho/ is mentioned by Nazarzoda, who also provides a variant in the AR script: انتبیناتکها [Nazarzoda et al. 2008: 1,70]. The latter is particularly interesting, as the word does not belong to the common DA-FA-TJ heritage, as it did not exist at the time, when the three idioms got isolated. Instead it is a part of a group of orthographic forms that where invented quite recently, when the discussion of reintroducing the AR script began. Consequently, it is remarkably different from the DA/FA forms.

The form TJ **антибиотик** may be used for further derivation, cf. the adj. **антибиотикӣ** [Muhabbat va Oila 2010-12-16] /antibyotik-i/.

We notice that most languages use the internationalism 'antibiotic' in some form, with FA having some alternative (neologisms) apart from that. In TR we find **antibiyotik** [Alkım et al. 1996: suppl., 4]. One could see that AR, like FA, has forms based on native elements: مضاد الجراثیم [Ba'albaki 1999: 52]; مُضَادٌّ حَیَوِيٌّ or مُضَادُّ الْحَیَوِیّاتِ and ضَنَادَّةٌ [Arslanyan & Šubov 1977: 44].

2.5 Artificial satellite

*The first artificial satellite, the Soviet **Спутник**-1 was launched in 1957 [Seeber 2003: 5].*

The form **радифи маснӯъ** /radif-i masnuʔ/ is to be found in Saymiddinov's dictionary [Saymiddinov et al. 2006: 336]. Interestingly, the plural form with the ending **-он** /-on/, i.e. **радифони маснӯъ** [Rahimov et al. 2006: 155] is attested, beside **-хо** /-ho/ [Radyo-i Ozodi 2009-02-13]. The results of an Internet search [Google search: keyword=радифи маснӯъ, domain=.tj, date=2010-11-17] are very low (about two hits), however, one has to remember that, just like in the case of the English term 'artificial satellite', **радифи маснӯъ** is an izofat phrase and where the context is clear enough, the word **радиф** [Myakišev & Buhovsev 2000: 137] may be used alone, or with some other modifier, e.g. **радифи телевизионӣ** /radif-i televizyoni/ 'a television [artificial] satellite' or **радифи обуҳавосанҷӣ** /radif-i obuhavosanǯi/ 'a meteorological [artificial] satellite' [See Osimi & Arzumanov 1985: 157]. On the other hand, a more elaborate variant **радифи маснӯъи Замин** [Normurod & Qodiri 2005: 108] or **радифи маснӯи Замин** (also pl. **радифхои маснӯи Замин** [Rahimov et al. 2006: 222]) is sometimes used, too. Moreover, the word **радиф** may be used in a number of different meanings, most of the not connected with space exploration at all [Saymiddinov et al. 2006: 488].

There is another form in TJ differing only in using a different (even if related and very similar, in fact) adjective, i.e. **маснӯъӣ** /masnuʔi/ (with an orthographical variant **маснӯй**), hence **радифи маснӯъии Замин** [Osimi & Arzumanov 1985: 1057] /radif-i masnuʔi-i zamin/ or **радифи маснӯй (Замин)** [Rahimov et al. 2006: 41]. The form is attested, albeit scarcely, in online resources [Radyo-i Ozodi 2009-02-13]. The antonym of these forms is **радифи табии Замин** [Habibullayev et al. 2010: 121] /radif-i tabi-i zamin/ 'the natural satellite of the Earth, the Moon'.

To the author's best knowledge, these forms are not used with the same meaning neither in FA nor in DA. The closest one is قمر مسنوعی /qamar-e masnuʔi/ attested both in FA [Klevcova 1982: 666; Āryānpur & Āryānpur 1375: 917; Haghshenas et al. 2002: 1504] and DA [Ostrovskiy 1987: 342; Fishstein & Ghaznavi 1979: 168]. A spoken form /qamar-e masnoye/ is also attested in DA [Awde et al. 2002: 58].

The forms for 'artificial' in the discussed lexical items are of AR origin (← <ṢNʕ> 'to produce'). In AR itself, the form for 'artificial satellite' uses a different form derived from the very same radix: ساتل اصطناعی [UNESCO acc. 2011-09-24].

Another form attested in a TJ publication issued in Uzbekistan is **хамсафари сунъии Замин** [Habibullayev et al. 2010: 118] /hamsafar-i sunʔi-i zamin/. It is similar to the ones discussed above, however, the form used for 'artificial' (an Arabism related to **маснӯъ**) is the one used in UZ, and specifically in the form: **сунъий йӯлдоши** [Koščanov et al. 1984: 498]. It may be found in a limited number of web publications [Radyo-i Ozodi 2004-04-13].

In most lexical works, the only meanings mentioned of the word **мохвора** /mohvora/ are those connected with the sense of 'a month' as a unit of time measure. E.g., according to Saymiddinov, it is either: 1. period of one month (= RU **месячник**), or it has an adverbial meaning: 2. 'every month' [Saymiddinov et al. 2006: 355]. The meaning 'artificial satellite' appears only in Nazarzoda's dictionary [Nazarzoda et al. 2008: 1,817]. However, various other (i.e. not lexicographical) sources use this form clearly in the context of space exploration [Radyo-i Ozodi 2010-02-05]. In online resources this form is used almost exclusively with the meaning of 'artificial satellite' and it is no doubt the most frequent TJ word for such an object [Google search: keyword=мохвора, date=2014-09-11]. It also appears with this meaning in a poem written in 1983 by Лоиқ Шерали, "*Чу моҳвора ба осмонҳо || кунем парвоз ба суръати нур...*" [Šerali [2009]]. **Моҳвора** is, no doubt, an Iranian word. Word-formationally it is a derivative of the noun **мох** 'the Moon', built with the FA/DA/TJ suffix **-vāre/-вора**. The original meaning of the form was adjectival, namely 'similar to the Moon, Moon-like' [Rubinčik 1970: 2,446]. On the basis of that, a nominal meaning has been developed 'a particular thing similar to the Moon → artificial satellite'. In FA a parallel form ماهواره is attested with the meaning of 'artificial satellite' both in lexicographical sources [Āryānpur & Āryānpur 1375: 917; Argāni 1364: 446] and in common usage (an Internet search within Iranian domains produces about 747,000 web pages containing this form, most of them in the meaning of 'artificial satellite' [Google search: keyword=ماهواره, domain=.ir, date=2010-11-20], which is a much better result than in the case of the rival form قمر مصنوعی, which is well attested in lexicographical works [Klevcova 1982: 666; Asadullaev & Peysikov 1965: 885; Āryānpur & Āryānpur 1375: 917] but brings less than 650 hits in an Internet search [Google search: keyword=قمر مصنوعی, domain=.ir, date=2010-11-20]). In DA, the same two forms known from FA are attested: قمر مصنوعی [Ostrovskiy 1987: 342] and ماهواره [Killid Group 1389-03-06 HŠ].

One should consider the possibility that – similarly to **хавопаймо** – the existing TJ lexeme **моҳвора** developed the new meaning under the influence of FA.

Further derivative from **моҳвора** is attested by Nazarzoda, namely the adj. **моҳвораӣ** /mohvora-i/, created using the suffix **-ӣ** /-i/ [Nazarzoda et al. 2008: 1,817]. A form with a variant of the same suffix: **-вӣ** /-vi/, i.e. **моҳворавӣ** /moh-vora-vi/ is attested, too [Prezident-i Toǧikiston [2005]; Tajik Air acc. 2011-03-27]. One instance of the same form but constructed with another variant of the same suffix: **-гӣ** /-gi/ has been found, i.e. **моҳворагӣ** [Google search: keyword=моҳворагӣ, date=2011-03-27][41] /mohvora-gi/.

Моҳвора is also used as a constituent of compounds, like in the case of the adj. **моҳворабар** /mohvora-bar/ 'carrying an artificial satellite' (e.g. in the phrase **мушаки моҳворабар** [Radyo-i Ozodi 2009-03-09; Malikov 2009] /mušak-i moh-vora-bar/ 'a rocket carrying an artificial satellite'), where the second element is the PrsS of the verb **бурдан** 'to carry' (See Fig. 7 below).

---

[41] The form has been found in the website http://www.shambari.kob.tj/namudi_hizmat.html. Unfortunately, by 2011-03-27 this page was inaccessible in its original form. It had been, however, cached by Google.

A RU loanword **спутник** /sputnik/ is used in TJ as well [Moukhtor et al. 2003: 234]. The plural of this form is attested in TJ as **спутникхо** [Kimyo-i saodat 2012-01-05] /sputnik-ho/. As far as the affixal derivation is concerned, the form **спутникӣ** /sputnik-i/ with the suffix **-ӣ** /-i/ is attested [Central Asian Voices acc. 2011-03-22].

At least Russian lexicographical works dedicated to the study of FA vocabulary mention a parallel form as attested also in this idiom: اسپوتنیک [Rubinčik 1970: 1,72]. However, analysing modern usage, it is hard to escape the conclusion that it is attested only as the rendering of the proper names like **Спутник**-1, **Спутник**-3 and not necessarily in the sense of a common term for 'an artificial satellite' [Google search: keyword=اسپوتنیک, date=2011-07-25].

As far as other languages of the region are concerned, in PŠ we find the form مصنوعى سپوږمى /masnuí spožməy/ [Lebedev 1961: 644]. DA, apart from the forms mentioned above, uses the EN loanword /satilāyt/ [Awde et al. 2002: 61].

2.6 Atomic bomb

The year when the first atomic bomb was constructed may be given precisely: 1945 [Encyclopaedia Britannica: vol. 1, 1990]. However, one has to be aware of the fact that as early as the 1934 the idea of atomic bomb was patented by Leo Szilard [L'Annunziata 2007: 246]. Moreover, the term 'atomic bomb' had been introduced even earlier in fiction literature, as it seems to have been used for the first time in a novel The World Set Free by Herbert George Wells [Michaelis 1962: 507], which was published in 1914. In the present chapter, apart from the 'atomic bomb' as such, we include terms that correspond rather to 'nuclear bomb' even if strictly speaking, the latter term has a broader meaning, designating also the 'hydrogen bomb'.

The form **бомбаи атомӣ** /bomba-i atomi/ is attested in lexicographical sources [Bertel's et al. 1954: 34; Osimi & Arzumanov 1985: 41; Nazarzoda et al. 2008: 1,228; Kalontarov 2008: 66], didactic literature [Habibullayev et al. 2010: 114] and an Internet search produces over 500 pages containing the form [Google search: keyword=бомбаи атомӣ, date=2010-11-21]. It is a stable *izofat* phrase, with the adjective **атомӣ** /atomi/ (derived with the abstract suffix **-ӣ** /-i/ from the noun **атом** /atom/) acting as an attributive modifier of the noun **бомба** /bomba/. It forms a regular plural, **бомбахои атомӣ** [Turažonzoda 2007-07-05; Mažidov & Nozimov 2006: 172].

The immediate source of the TJ word **атом** is most probably the parallel RU form. The elements of this phrase are used in many languages. The original source of the internationalism **atom** is the GR form **ἄτομος**. It was borrowed into LA as **atomus** and then it was passed on to numerous languages of Western Europe, whence it came to RU (**атом**), where is well attested in the 18th century [Černyx 1991: 1,58]. It seems most probable that RU served as a vehicular language in passing the form further into TJ.

The GR βόμβος 'a clunk, crash, noise, etc.' [Groves 1844: 116] is the source of the LA **bombus** 'a noise, buzz'. From the latter come Italian **bomba** 'a bomb' and the French **bombe** with the same meaning [Černyx 1999: 103]. In RU the word has been present since the reign of Peter the Great, however, it is difficult to indicate the vehicular language, via which it came to Russia [Černyx 1999: 1,103]. Preobraženskiy believed it might have been borrowed either directly form FR or via DE [Preobraženskiy 1958: 36]. The word was introduced into TJ in the form بومبه in the pre-Soviet period [Rzehak 2001: 137].

There are parallel phrases in both FA and DA to the TJ **бомбаи атомӣ**, however, in each case, the head has a different form owing to a specific route on its way to the target language. In the TJ بومبه / **бомба** the final **-a** indicates RU as the immediate source. The FA بمب /bomb/ in بمب اتمی [Klevcova 1982: 46; Asadullaev & Peysikov 1965: 63; Āryānpur & Āryānpur 1375: 187] seems to indicate FR as the vehicular language. One should note the difference between the TJ form and the FA form, namely the FA variant بمب /bomb/ seems to be borrowed from the FR **bombe**. The phrase بمب اتمی is attested in DA, as well, however, there is also a specifically DA form بم اتمی /bam-e atōmī/[42] [Lebedev 1989: 52; Google search: keyword=بم اتمی, date=2014-03-22], where we find بم /bam/ [Lebedev 1989: 31, 52] most probably borrowed from PŠ, as many other military terms [Dorofeeva 1960: 65]. The source of the latter, in its turn, is the spoken EN /bɒm/.

Interestingly, a derussianized variant of **бомбаи атомӣ** that is **бомби атомӣ** / bomb-i atomi/ exists in TJ, too, however, it has been found only in online publications [Radyo-i Ozodi 2011-06-23; Jumhuriyat acc. 2011-07-09]. It must have been influenced in recent times by the FA form بمب /bomb/ (even though the traditional vowel correspondence rules are violated here) and/or the EN orthography [Gacek 2014: 152-153].

Some other languages of the region compose parallel forms using same elements but implying syntactical mechanisms typical of them. E.g. in PŠ we find phrases د اتوم بم /də aṭóm bam/ [Lebedev 1961: 31] and اتومی بم [Lebedev 1989: 52]. In UZ we find **атом бомбаси** [Koščanov et al. 1983: 81] and – similarly – in TR/AZ: **atom bombası** [Media forum 2011-01-18; Alkım et al. 1996: 94]. These show, again, RU influence, while the construction of the phrase is that of the so called Turkish izafet. Finally, in KY we find **атомдук бомба** [Yudahin 1957: 31].

Another TJ form, the phrase **бомбаи хастаӣ** /bomba-i hastai/ is attested in lexicographic works [Nazarzoda et al. 2008: 2,499] and in online sources (an Internet search results in 100 websites [Google search: keyword=бомбаи хастаӣ, date=2010-11-26; cf. Mirzob 2010-05-27; Muhabbat va Oila 2010-03-05]). It is simillar in its structure to **бомбаи атомӣ**, however, it has a native adjective **хастаӣ** (derived from the noun **хаста** 'nucleus') as a modifier. The form **бомбаи хаставӣ** /bomba-i hasta-

---

[42]   The form بمب with pronunciation /bamb/ is attested as well [Awde et al. 2002: 30].

vi/ differs from the previous one only in applying the another form of the same adjective, **хаставӣ** created with the variant of the adjectival suffix /-i/, i.e. /-vi/[43].

Interestingly, as far as the online sources are concerned, the form **бомбаи хаставӣ** has been found only in **Радиои Озодӣ** pages and citations from there [Google search: keyword=бомбаи хаставӣ, date=2010-11-26]. However, a related form **силоҳи хаставӣ** /siloh-i hastavi/ is attested in printed texts [Muzofiršoyev 2009: 46].

The corresponding form بمب هستهای is attested both in FA [Google search: keyword=بمب هستهای, domain=.ir, date=2010-11-26] (over 44,000 results) and DA [Google search: keyword=بمب هستهای, domain=.af, date=2010-11-26] (over 50 results). The phrase بمب هستوی has been found only in Afghan (supposedly DA) sites [Google search: keyword=بمب هستوی, date=2010-11-26]. This seems to be more or less true about the form هستوی itself, too (The results of searching this form within Iranian domains are low, below 400 hits [Google search: keyword=هستوی, domain=.ir, date=2010-11-26]), especially when compared to over 2,500,000 hits in the case of هستهای [Google search: keyword=هستهای, domain=.ir, 2010-11-26]. In other words, هستهای seems to be a form typical for eastern types of Persian (be it DA or TJ)[44].

Moreover, both variants of the adjective in question seem to have been borrowed into PŠ, as هستئی /hastayí/ and هستوی /hastawí/ [Aslanov 1966: 971]. Hence, 'nuclear bomb' happens to be called د هستوی بم /də hastawí bam/ [Google search: keyword="د هستوی بم", date=2010-11-26].

Returning to the sphere of the Tajik lexica, we note that just like in the case of the pair **бомбаи атомӣ : бомби атомӣ**, there exists a form related to **бомбаи хастай** differing only in using the de-russianized variant of the noun in question, i.e. **бомби хастай** [Salimpur 2009].

The form **бомбаи ядрой** /bomba-i yadroi/ has both structure and semantics parallel to those of **бомбаи хастай (хаставӣ)**. The difference between the two is that **ядрой**[45] is an adjective derived (again using the suffix /-i/) from the RU loanword **ядро** 'nucleus' (itself used in TJ [Saymiddinov et al. 2006: 764]) – an equivalent of FA/TJ /haste, hasta/.

Below 5 instances of **бомбаи ядрой** in online sources have been found [Google search: keyword="бомбаи ядрой", date=2010-11-26]. A similar form: **силоҳи ядрой** is attested in lexicographical sources [Saymiddinov et al. 2006: 764]. No de-russianized form like ***бомби ядрой** has been found.

---

[43] Both morphonological variants of the suffix /-i/, namely /-gi/ and /-vi/ are often used in modern TJ not consistently that is against the etymological factor which should be, in theory, decisive in choosing one of them [c.f. Perry 2006: 426].

[44] Actually, the use of /-vi/ instead of /-i/ is quite restricted in FA. Interestingly, apart from some forms of AR origin, /-vi/ it happens to be attached to certain toponyms of the eastern extremities of the Persian speaking region, like هرات 'city of Herat' → هروی /heraví/, دهلی /dehli/ 'city of Delhi' → دهلوی /dehlaví/.

[45] The AR script variant is یدرایی [Nazarzoda et al. 2008: 2,680].

As we see, all the forms discussed are based on at least one loanword (the word from 'bomb'). In some other languages native lexical items for 'atomic bomb' may be found, e.g. AR القنبلة الذرية [Baʿalbaki 1999: 71].

2.7 Automated Teller Machine (ATM)

Establishing a non-ante date in the case of ATM is not an easy task. Theoretically, everything is clear: L. Simjian patented a crude, rudimentary variant of this device in 1939 in the US, so this is the date we use in the classification in the present work. However, one has to remember that Simjian's invention met no real public interest, so neither the invention nor its name got popularized by then. The next try was made in 1962, in the US, too. Still, it was hardly a success. It was only another type of ATM designed in 1973 by a team of the Docutel Corporation that gained some popularity [Reilly 2003: 23].

The word **банкомат** /bankomat/, to the author's best knowledge, is not yet attested in the lexicographical sources. However, it may be found e.g. in TJ press and in the modern electronic sources [Agroinvestbonk 2011-06-15; Mahmadbekova 2011-01-14; Quqanšoh, 2011-01-19]. It is an internationalism present in a number of languages, RU & TJ included.

Some sources testify that the ATMs are quite a new item in the Tajikistan's landscape [Quqanšoh 2011-01-19] and so is the term for them. Some details of the form's orthography may possibly indicate that it is perceived as not entirely assimilated by the language users, e.g. its plural form (which is, by the way, well attested) is sometimes written with the stem in guillemets and the plural ending attached by hyphen **«банкомат»-хо** [Amonatbonk acc. 2011-01-31]. We have to note, however, that it is by no means a common practice (cf. **банкоматхо** /bankomat-ho/[46] [Oriyonbonk 2009; BMT 2011-01-12; etc.]). Hyphenated in writing or no, nevertheless, existance of the plural form supports the idea that **банкомат** is to some extent nati vised in TJ. The form is in fact undergoing that process right now. The author had been unable to find any derivatives of this form before the middle of the 2011 year. However, from this time, at least the form with the adjectival suffix **-й** /-i/, i.e. **банкоматй** / bankomat-i/, has been attested [Xoliqzod 2011].

The EN acronym **ATM** seems to be used in TJ, as well, at least where an abbreviation is needed [Amonatbonk acc. 2011-01-31]. Interestingly, an example of adding plural suffix **-хо** to this acronym has been found: **atm-хо**. This form appears in a message posted by certain Кактус on 2010-07-26 16:53 as a comment to an article published by **Радиои Озодй** [Radyo-i Ozodi 2010-07-26]. Nevertheless, it is not assimilated graphemically, so we shall classify it as a foreign word.

---

[46] Pronunciation attested [Radyo-i Ozodi 2012-02-07].

The form **дастгохи худпардоз** /dastgoh-i xudpardoz/, is used as well and it is attested in a number of online resources. It is a parallel to the FA where we find, دستگاه خودپرداز [Bānk-e Sāderāt-e Yazd 1388-02-05 HŠ]. Existence of this form in FA and the fact the atms appeared in Iran earlier than in Tajikistan entitles us to put forward the hypothesis that **дастгохи худпардоз** is a loanword from FA.

Apart from the izofat phrase **дастгохи худпардоз**, also the compound **худпардоз** alone may be used in the sense of ATM [TSB acc. 2014-06-22; Dariush 2008-02-24], however, this word may be also used in other senses [Nazarzoda et al. 2008: 2,27].

Apart from the ones mentioned above, FA uses different forms, in particular عابربانک /āberbānk/ [Tabnak 1390-03-31 HŠ]. Two more FA counterparts have not been found in the lexicographic works, being however quite popular in electronic publications: خودپرداز /xodpardāz/ (with an Internet search producing over 162000 results [Google search: keyword=خودپرداز, date=2011-01-31]) and زودپرداز /zudpardāz/ (below 400 results [Google search: keyword=زودپرداز, date=2011-01-31]) – both compounds based on native elements. The author of the present work has found no trace of these forms in TJ.

2.8 Automatic wind-shield wipers

Automatic wind-shield wipers were invented in 1921 (manually operated ones preceded them by almost twenty years) [Baron & Shane, 2007, 237].

The TJ name for wind-shield wipers is **барфпоккун** برفپاککن [Nazarzoda et al. 2008: 1,148] /barf-pok-kun/. Online search results for this form are quite low (below 10 results) [Google search: keyword=барфпоккун, date=2014-09-11].

It is identical with the FA form برفپاککن [Argāni 1364: 553]. From the word-formational point of view, this form belongs to a quite popular type of FA/DA/TJ compound words (cf. مرا فراموشمکن 'forget-me-not', lit. 'do not forget me!') containing a predicative element, formed by means of lexicalisation of a sentence [Rubinčik 2001: 140]. They seem to find a parallel in forms like the EN **merry-go-round**, **forget-me-not** and FR **ne m'oubliez pas**.

The problem to solve is, whether the form in question appeared first in FA or TJ. Anyway, **барфпоккун** seems to be an original Iranian form, not a borrowing or a calque from some other language. Were we to put forward the hypothesis about the original source of the form, we would point rather at FA, than TJ. An important reason for this is that it is in most cases possible to indicate the more typical direction of lexical borrowings between two idioms (e.g. much more often AR → FA, than FA → AR; rather EN → FA, than the opposite, etc.). In the case of FA and TJ, the typical direction is certainly FA → TJ, as it is attested by the documented examples like هواپیما → **ҳавопаймо** (see ch. 2.2), رایانه → **роёна**. Thus, as long as no contradictory data is found, we will assume that the FA برفپاککن was reflected in TJ as **барфпоккун** and not vice-versa.

Also, a shorter form is used in TJ **барфпок** برفپاک [Nazarzoda et al. 2008: 1,148] /barf-pok/. It may be found in electronic publications, however, it is not much more popular than the previous form [Google search: keyword=барфпок, date=2014-09-11].

A parallel form is used in DA: /barf-pāk/ [Awde et al. 2002: 210] and in FA (where it seems to be quite rare) [Google seach keyword=برفپاکها, date=2012-10--29]. This seems to be a result of clipping (cf. EN **telephone** → **phone**).

The RU **стеклоочиститель** seems to be used on a limited scale, possibly as a foreign word [Šarifov acc. 2014-09-11].

2.9 Black box

Providing one single date of invention of flight recorders popularly known as black boxes is not an easy task. Primitive forms of flight recorders were used from the beginnings of aviation, including one applied by Wright brothers themselves [Volti, 1999, 413]. These were, of course, very simple devices, recording only the few most important flight parameters. However, if we speak of black boxes nowadays, we mean devices that have to possess certain features: they are placed in a plane to help investigating incidents and accidents, esp. when there are no survivors; they have to be capable of recording most flight parameters, they have to be crash-proof. They also have to record cockpit voices (even if these functions are performed by more than one specialistic devices, practically they are all referred to collectively as black box), they should be traceable if lost e.g. in the ocean.

The form **чаъбаи сиёх** is used in TJ (especially in online resources) [BBC Persian 2009-07-16; Radyo-i Ozodi 2006-08-23] /ʒaʔba-i siyoh/ and in FA: جعبهٔ سیاه [Haghshenas et al. 2002: 125]. It is an izofat phrase with the noun **чаъба** 'box' (← AR) as the head, and the native adj. **сиёх** 'black' as the modifier.

Taking into consideration that TJ uses a form identical to that of FA we may put forward the hypothesis that we are dealing here with an example of lexical borrowing between the two closely related idioms, like in the case of the form **роёна** (q.v.). However, while the example of **роёна** is clear, here additional research is needed to verify the hypothesis. The primary source is most probably the EN **black box** of which the FA form is a calque. However, in the author's opinion, one may not exclude the possibility that the TJ **чаъбаи сиёх** may be a calque, independent from FA, based directly on the RU **чёрный ящик**, as the lexemes used in the TJ form seem to be the natural counterparts of the elements of the RU phrase.

As far as the original source is concerned, it seems quite possible that a form parallel to the EN 'black box' was first used in FR (**boîte noire**) by François Hussenot referring to the device he constructed around 1960. Its interior had to be painted black, as the data were registered photographically [Beaudoin, 2005, 210]. Taking this into consideration, one could present the EN **black box** as a calque from FR.

Nowadays, the borrowing from EN, **black box** is used in FR and in DE (**Blackbox** beside a calque **schwarzer Kasten** [Želyazova 2005: 92]. In RU we find **чёрный ящик** [Želyazova 2005: 92], however, it is difficult to indicate the direct model for this calque. It may be the EN form, however, other sources (esp. the original FR and DE forms) cannot be excluded.

2.10 Blood type

Blood types (or Blood Groups) were discovered by K. Landsteiner in 1901 [Brze-ziński, 1995, 355].

The form **гурӯхи хун** [Ustoyev 2008: 56; Maǯlis-i Oli 2005-03-01; Reporter.Tj 2009-06-04] /guruh-i xun/ is used in TJ, including electronic publications. Structurally, **гурӯхи хун** is a stable izofat phrase. The plural form **гурӯххои хун** [Ustoyev 2008: 56] /guruh-ho-i xun/ is used, too.

The concept of blood groups has been introduced by Karl Landsteiner in his publications starting in 1901 [Landsteiner 1901]. These were written in German, so we may assume that the DE **Blutgruppe** is the primary source of the form. Numerous languages introduced calques of this form (c.f. EN **blood group**, FR **groupe sanguin**, RU **группа крови** etc.). Thus, it is difficult to indicate the particular source of both RU and TJ forms. We may only suspect that the RU form is a calque of some of the Western European terms (esp. DE, EN or FR), while the TJ one might have been coined on the basis of the RU one. However, we may not exclude a priori the possibility that the form was borrowed from FA, where a parallel form گروه خون is attested [see Google search details below].

Another – closely related – TJ expression conveying the meaning of 'blood type' is **гурӯхи хунӣ** /guruh-i xuni/. It has been found, however, only in one sole website [BBC Persian 2012-11-14]. The difference between the previously commented phrase and this one is that here the adjective **хунӣ** 'of blood, concerning blood; bloody' is used as the modifier instead of the noun **хун**. Again, we find a parallel form in FA: گروه خونی [Haghshenas et al. 2002: 132]. In fact گروه خونی is the dominating form for 'blood group' in FA, with the frequency much bigger than گروه خون [Google search: keyword=گروه خون, date=2014-05-14; Google search: keyword=گروه خونی, date=2014-05-14]. A similar izofat phrase is attested in DA, as well: /gorup-e xuni/ [Awde et al. 2002: 40]. Here, however, the loanword /gorup/ (with EN as the immediate source) is used instead of the native form. The phrase with a noun as the modifier, i.e. /gorup-e xun/ [Awde et al. 2002: 74] is used, too, in DA.

In TR we find **kan grupu** [Alkım et al. 1996: 592].

2.11 Bluetooth technology

The Bluetooth technology was invented by ERICSSON in 1994 and named after the king Bluetooth II [Malaga, 2005, 134] (Harald Blåtand II).

The form **Bluetooth** is used in TJ and, interestingly, it is often written with Latin letters [NOKIA 2010-07-19: 39], even if sometimes the Cyrillic form **Блутус** is used, too [Ozodagon 2011-05-05]. Both forms show no particular features either supporting or contradicting the possibility of RU playing the role of a vehicular language in this case. Use of either Cyrillic or Latin script is not a decisive factor here.

The author was unable to find either plural or the expected derivatives (***Блутусдор**, ***беблутус**, etc.). Thus, both **Bluetooth** and **Блутус** will be classified rather as foreign words than loanwords (**Bluetooth** in particular, as it is not assimilated graphemically, too).

A similar form is used in FA and AR, as well, and the orthography is identical: بلوتوث. This is interesting particularly in the case of FA, as it makes it one of the examples showing that even though the standard Persian pronunciation of the letter ث is simply /s/, still there is a tradition to render the foreign interdental spirant by this letter (just like in AR, where it is justified by the native pronunciation). This may be confronted with the forms like the one used in Kurdish: بلوتوس.

The EN **Bluetooth** seems to be a calque of the DN **Blåtand** with, additionally, alternation of meaning. It is no longer a personal name. Then Bluetooth seems to be borrowed by TJ.

Another TJ form attested in online resources is the calque of the internationalism in question **Дандони обй** [Faryod-i xomŭš 2011-05-05]. Parallel form may be found in other languages, e.g. in UR: نیلادانت [Wikipedia: entry=نیلادانت, date=2011-10-09] /nīldāṅt/ (a not very popular calque used beside the EN form). It is not clear whether the calque **Дандони обй** is modelled directly upon the DN origin, the EN form or rather the RU **синий зуб** which is not very popular, but – nevertheless – existent.

2.12 CD-ROM

CD-ROMs were invented in 1984 [Chadwyck-Healey, 2011, 456].

The form **компакт-диск** /kompakt-disk/ together with its plural **компакт-дискхо** /kompakt-disk-ho/ is attested in TJ [Komilov & Šarapov 2003: 63ff.]. Identical form is used in RU, so we may put forward the hypothesis that this internationalism of EN origin was transferred into TJ via RU.

The internationally used acronyms **CD** [sidí] and **CD-ROM** (← EN Compact Disk-Read Only Memory)[47] are used in TJ, too [Komilov & Šarapov 2003: 64; Radyo-i Ozodi 2006-04-06]. These forms, especially **CD,** may also be used in plural [Radyo-i Ozodi 2006-04-06]. Both forms are of EN origin. We may suspect RU as an intermediary in their transfer into TJ. All these forms are not graphically assimilated in TJ, therefore they should be classified as foreign words.

Interestingly, the above-mentioned acronyms are used in numerous languages originally written with non-Latin writing systems, see e.g. Japanese **CD** [šīdī] [Kay 1995: 70], e.g. in the phrase ブータブルＣＤ /būtaburu-šīdī/ 'a bootable CD'.

In FA the borrowed acronym سیدی is used together with the form دیسک فشرده [Haghshenas et al. 2002: 205]. The DA spoken form is /si-di/ [Awde et al. 2002: 35].

2.13 cell phone (mobile phone)

The work on the cell phone technology started in the 1940s. Ericsson in Sweden and NTT in Japan experimented with this idea providing some kinds of cell phone networks from 1959. Nevertheless, cell phones as we know them today (i.e. with multiple non-interfering channels 'packed' into one frequency, automatic switching from one base station to another, while the user is moving between the virtual cells etc.) may be traced to 1973. This is when Martin Cooper of Motorola, commonly known as the creator of cell phone, presented his invention. Otherwise, one may stick to 1979, when the patent for such technology was granted to C. A. Gladden and M. A. Parelman [Klooster 2009: 416]. Discussing this item, one has to pay attention to some subtleties. First of all, mobile phone is not necessarily a cellular phone. Cellular technology is just one of the ways to supply the public with mobile telephone sets. On the other hand, cellular phones are, no doubt, a dominating type of mobile phones used today, so the term 'mobile phone' often refers practically to cell phones.

The form **телефони ҳамроҳ** /telefon-i harmoh/ is well attested in the electronic sources – an Internet search produces nearly 5000 results [Google search: keyword="телефони ҳамроҳ", date=2011-02-10]. This form is used in particular by mass-media [BBC Persian 2010-01-07; Millat 2009-02-04; Radyo-i Ozodi 2009-03--10; etc.]. As far as the structure of this form is concerned, it is a stable izofat phrase, with the loanword **телефон** /telefon/ as the head, and the modifier **ҳамроҳ** /hamroh/ 'accompanying'. The word **telephone** was coined in EN of GR elements: τῆλε 'far' [Ewing 1827: 827] and φωνὴ 'voice, sound, etc.' [Groves 1834: 597; Haspelmath & Tadmor 2009: 369; Eddy 2007: 55-56]. It seems possible that the EN **telephone** was borrowed directly into RU [Eddy 2007: 82], from where it was transferred into TJ: **telefon** [Rzehak 2001: 348] in the Cyrillic script: **телефон**.

---

[47] To be precise, these forms are not synonyms. However, they are used as such in everyday language.

The parallel FA form is well attested, too [Haghshenas et al. 2002: 1081; Google search: keyword=تلفون همراه, date=2011-02-10].

The forms **телефони мобайл** [Radyo-i Ozodi 2009-02-20; BBC Persian 2008-06-05] /telefon-i mobayl/, **телефони мобил** [Radyo-i Ozodi 2012-08-09; Qadamova 2010-04-19] /telefon-i mobil/ and **телефони мобилӣ** [Khovar 2008-08-21; Radyo-i Ozodi 2010-07-27] /telefon-i mobili/ all belong to the same family. They are izofat phrases with **телефон** /telefon/ as the head. The modifier is either loanword from EN: **мобайл** /mobayl/[48], or **мобил** /mobil/, which might have been taken from RU and deprived of the original adj. suffix. The form **мобилӣ** /mobil-i/ presents the next level of adaptation – the loanword **мобил** /mobil/ gained the native adjectival suffix **-ӣ** /-i/ (cf. forms like **аллергӣ**, where a TJ suffix is added to an arbitrary stem of the RU form). The RU form **мобильный** comes from FR **mobile**, in its turn derived from the LA **mobilis** [Spirkin et al. 1980: 325]. The EN **mobile** may be traced back to the very same FR **mobile** [Skeat 1993: 296].

Related forms, e.g. تلفن موبايل /telefon-e mobāyl/ or simply موبايل /mobāyl/ do exist in FA, as well [Google search: keyword=تلفن موبايل, date=2011-02-11]. In DA we find /mobāyl/ [Awde et al. 2002: 94], too. The same elements are used in the equivalents in the other languages of the region, e.g. UZ **mobil telefon** (besides forms like **uyali telefon** and **qo'l telefoni**) [Wikipedia: entry=mobil telefon, date=2011-02-11].

An interesting form is **телефони дастӣ** [Ayubzod 2012-04-25; Hizb-i Nahzat 2011-12-20] /telefon-i dasti/. It is, just like **телефони мобайл** and the related forms an izofat phrase with the loanword **телефон** as the described part. The attributive element in this case is a native adj. **дастӣ** /dast-i/, created by adding the suffix **-ӣ** /-i/ to the noun **даст** /dast/ 'hand'. This form is attested in FA, too [Google search: keyword=تلفن دستى, date=2011-02-11].

Now, of all the TJ forms above, we would say that **телефони хамрох** and **телефони дастӣ** may be borrowings from FA (for the reasons discussed while analysing **барфпоккун** – see ch. 2.8). On the other hand, **телефони мобайл**, **телефони мобил** and **телефони мобилӣ** seem to present examples of combined instances of lexical borrowings (both elements borrowed at different times) and calque, as far as the structure of these forms is concerned.

2.14 Cellophane

Cellophane was invented in 1908 by J. E. Brandenberger [Hall, 2008, 25].

The internationalism **селлофан** /sellofan/ is used in TJ, both in lexicographical works [Saymiddinov et al. 2006: 536; Moukhtor et al. 2003: 225] and online resources [Farmon 2013-05-25]. Alternative pronunciation: /salafan/ is attested for

---

[48] The diphthong in the second syllable is an indication that the form has been borrowed from EN.

Bukharan TJ [Ido 2007: 12]. A derived adjective **селлофанӣ** [Ozodagon 2011-07-22; Botanix 2009-06-06] /sellofani/ is used in TJ, as well.

The [FR] word **cellophane** was coined by the inventor of this material of the FR words **cellulose** and **diaphane** [Hounshell & Smith 1988: 170]. **Cellulose**, in its turn, was created by Anselme Payen on the basis of the FR **cellule** 'living cell' with the addition of the suffix **-ose** [Pérez & Samain 2010: 33; Senning 2007: 72]. The form **cellule** comes from the New LA **cellula** 'living cell' which is technically a diminutive of the LA **cella** [Senning 2007: 72]. The FR **diaphane** is derived from the GR **διαφανὴς** 'clear, transparent, pellucid' [Senning 2007: 72; Groves 1834: 148]. It seems that in RU the form **целлофан** might have been borrowed via DE (or DE and PL), as it is suggested by the initial affricate. The very same affricate got lost again when the form was transferred from RU to TJ, as in the case of **цемент** → **семент** [Sanginov 2010: 43], **цирк** → **сирк** [Sanginov 2010: 47], etc.

In FA سلوفان [Asadullaev & Peysikov 1965: 1011; Haghshenas et al. 2002: 206] is used. Apart from that, the phrase کاغذ طلق is attested, too [Google search: keyword="کاغذ طلق", date=2015-01-28]. According to Ostrovskiy, the DA equivalent is پلاستیک /palāstik/ (sic!) [Ostrovskiy 1987: 381].

In UZ [Koščanov et al. 1984: 703] and KY [Yudahin 1957: 938] similar forms are used: **целлофан** and in AR we find السيلوفان [Ba'albaki 1999: 162].

2.15 Cluster bomb

Air-dropped cluster munitions were first used during the World War II in 1943 by Soviet and German forces [Human Rights Watch 2007: 8].

The TJ form **бомбаи кластерӣ** /bomba-i klasteri/ (often in plural – **бомбаҳои кластерӣ** /bomba-ho-i klasteri/) is attested in only a limited number of sources [Mažalla-i Bomdodi, 2010-07-15; ICBL, 2009]. As far as the structure of the form is concerned, it is an izofat phrase containing two loanwords, one of them with a native adjectival suffix **-ӣ** /-i/ added (see Fig. 8).

The phrase **бомбаи кластерӣ** seems to be a calque of either EN **cluster bomb** or RU **кластерная бомба** (on the origin of the form **бомба** /bomba/ see discussion on **бомбаи атомӣ**, p. 46). The word **кластер** has been most probably borrowed from EN **cluster** and RU played the role of the vehicular language in transferring this word to TJ. It is used in TJ in other contexts, as well, just like its derivative built with the adjectival suffix **-ӣ**.

No trace of use of this phrase in FA or DA has been found, however, a direct loanword form EN based on the very same elements is quite frequent: کلاستر بمب / kelāster-bomb/ [Google search: keyword=کلاستر بمب, date=2011-02-14].

Apart from the above-mentioned form, the phrases **бомбаи кассетӣ** /bomba-i kasseti/ (more often in pl., **бомбаҳои кассетӣ** /bomba-ho-i kasseti/) and **бомбаи кассетавӣ** (pl. **бомбаҳои кассетавӣ** /bomba-ho-i kassetavi/) are used in TJ [Maǯalla-i Bomdodi, 2010-07-15]. They are structurally similar to **бомбаи кластерӣ** and – at the same time – related to each other. Instead of the adj. **кластерӣ** we have either **кассетӣ** or **кассетавӣ** here. They are both related to the internationalism 'cassette', the original source of which is the FR **cassette** [Spirkin et al. 1980: 224]. However, the latter shows the RU influence here, as the final /-a/ of the **кассета** /kasseta/ is clearly a RU feminine declension marker. This ending of the form **кассета** induces the choice of the special variant of the adjectival suffix **-ӣ** /-i/, i.e. **-вӣ** /-vi/. The whole form is probably a calque of the RU **кластерная бомба**. The form **кассетӣ** is also used in other contexts (Saymiddinov et al. 2006: 272). On the other hand, the alternative form **кассетавӣ** does not seem to be used apart from the phrase **бомбаи кассетавӣ**.

Finally, there is also a form **бомбаи хӯшай** (pl. **бомбахои хӯшай** /bomba-i xŭšai/) [Maǯalla-i Bomdodi 2010-07-15], which is quite similar in structure to the previously discussed forms. The only difference is that the modifier is the native adj. **хӯшай** /xŭšai/, derived from the noun **хӯша** /xŭša/ 'cluster, bunch' by the addition of the suffix **-ӣ** /-i/. A parallel form is attested in FA: بمب خوشه‌ای /bomb-e xuše-i/ and in DA: /bam-e xošayi/ [Awde et al. 2002: 205], so we may not exclude the possibility that it is a borrowing from a closely related idiom.

2.16 Computer

*'Computer' is one of the words that cannot be omitted when discussing the development of science and technology of the 20th century. Nevertheless, it is not so obvious, what is understood under that term. Some tools used for computing are thousand years old (e.g. the famous Sumerian abacus). Certain mechanical devices developed in the 19th century were capable of at least a part of functions modern computers are expected to fulfil (acting in accordance with some form of a programme, recording data on some medium, performing mathematical calculations). However, the author of the present work believes that the word **computer** is used today only when referring to devices that may be characterised as being: a) digital (not analogue), b) electronic (rather than mechanical or electro-mechanical), c) programmable (some calculators possess this feature, too), d) universal (they may be used for various, unlimited purposes, as opposed to mobile phones, calculators, etc.)[49], d) capable of communicating with various peripheral devices. The first device that could be described like that was probably the American ENIAC of 1946 [see Reilly, 2003, 91][50].*

---

[49] This criterion seems to be most problematic nowadays, when we are surrounded by devices, which are difficult to classify, e.g. smart-phones.

[50] The German Z3 was electromechanical, not electronic. The British Colossus computers and the American Atanasoff–Berry Computer were not (as opposed to ENIAC) multi-purpose.

Probably the most widely used TJ term for 'computer' is the very same internationalism appearing in at least two variants: **компютер** [Saymiddinov et al. 2006: 280; Baizoyev & Hayward 2003: 339; Satskaya et al. 2007: 90; VKD 2014-01-08] /kompyuter/ [Ido 2007: 25] and **компутер** [Moukhtor et al., 2003, 103; Abdulov 2010-02-17; Ruhulloh S. 2012-07-01] /komputer/. As far as the first one is concerned, its form clearly indicates RU as the vehicular language. In the other case it is not so easy to determine the intermediary.

Derivatives and compounds based on these forms are attested, too, e.g.: **компютери** [Saymiddinov et al. 2006: 280; Komilov & Šarapov 2005: *passim*] /kompyuter-i/ or **компютеркуни** 'computerization' [Bashiri, 1994, 120][51] / kompyuter-kun-i/.

Similar forms are used in other languages, too: FA کامپیوتر [Gacek 2007: 19], DA کمپیوتر /kampyutar/ [Ostrovskiy 1987: 155], /kāmpyutar/ [Awde et al. 2002: 44][52], UZ **kompyuter** [Balci et al. 2004: 130].

The form **роёна** is not as popular as the previous forms but, nevertheless, it is attested [BBC Persian 2004-05-17; Tožikon acc. 2010-10-06]. It seems to be a recent borrowing from FA, where this form (رایانه) is used [Gacek 2007: 24]. The derived adjective [Radyo-i Ozodi 2007-11-21] /royona-i/ is attested, too.

The FA form رایانه – in its turn – is a quite recent neologism. It is still absent from the sources of 1980's [see Argāni 1364: 116].

2.17 Computer file

Punched cards were grouped in units called 'files' as early as the 1950s.

TJ uses the internationalism **файл** [Nazarzoda et al. 2008: 2,381; Islomov et al. 2005: 4ff.; Komilov & Šarapov 2005: 6ff.] /fayl/ (the AR script فیل [Nazarzoda et al. 2008: 2,381]), the original source of which is EN (the form itself has a long history going back at least to LA, however, in reference to a portion of computer data it was first used in EN). The same form is used in RU, too, so the latter may have served as an intermediary.

Apart from lexicographical works, the form is attested in online resources, as well [Oftob acc. 2014-09-11; Tcell acc. 2014-09-11]. A plural form **файлхо** /fayl-ho/ is to be found, too [Komilov & Šarapov 2005: 31; Islomov et al. 2005: 5; Google search: keyword=файлхо, date=2011-03-23]. A derived adjective **файли** /fayl-i/ is attested, too [Komilov & Šarapov 2003: 92; IPD GBAO acc. 2011-03-23].

Just like in the case of many other internationalisms, we are entitled to put forward the hypothesis that RU served as the vehicular language in transferring this

---

[51] Bashiri translates this form as 'to computerize', this, however, is obviously a mistake, as **компютеркуни** is clearly an abstract noun and not a verb.

[52] An interesting form for 'a laptop' is attested in DA: It is the izofat phrase /kāmpyutar-e syār/ [Awde et al. 2002: 44].

word from EN into TJ. It is in EN that the form acquired the additional meaning of 'a set of computer data' on the basis of the analogy to the pre-existing meaning 'a container for documents'. The word **file** in EN is a borrowing from FR **file**, in its turn taken from the LA **filum** 'thread' [see Skeat 1993: 130].

FA uses variants of the same internationalism: فایل /fāyl/ [DTM 1391-09-06] and فیل /fayl/ [Gacek 2007: 20], apart from a calque پرونده /parvande/ [Gacek 2007: 20], of which the first one seems to be the most popular (over 1,000,000 results in an Internet search [Google search: keywords=فایل + رایانه, date=2010-12-25][53]), the second is slightly less attested (over 900,000 results [Google search: keywords=رایانه + فیل, date=2010-12-25]) and the last is relatively seldom used (below 200,000 hits [Google search: keywords= پرونده + رایانه, date=2010-12-25]) even if promoted officially as a 'native form'.

In DA the form /dosiya/ is used, just like in the case of paper files [Awde et al. 2002: 177]. Similar forms are used in other languages of the region, e.g. UZ **fayl** [Kuvatov 2010-11-29].

2.18 Computer memory

What seems to be the first type of computer memory is the 'drum memory' invented by the Austrian inventor Gustav Tauschek in 1932 [Bachman, 2010, 214]. It was preceded by some earlier ways of storing data (like punched cards applied first in the inventions of Charles Babbage), however, these were – in the author's opinion – related rather to modern external storage devices, rather than computer memory sensu proprio.

The word **хофиза** /hofiza/ [Gacek 2007: 24] is an Arabism. However, it has been present in the FA/TJ/DA for centuries with the meaning of 'memory' in general, to acquire the meaning of 'computer memory' only in the last decades. Not surprisingly, the same happened to the parallel FA form حافظه /hāfeze/ [Gacek 2007: 24]. This example illustrates that although – in general – the period of massive borrowing of Arabic words into DA/FA/TJ is over, still forms of AR origin may be found in the newest levels of the TJ lexica. In FA, apart from حافظه, a number of other words is used in the sense of 'memory' in the computers' context: یادانبار [Google search: keyword="یادانبار", date=2014-10-12] and یادگنجایی [Google search: keyword="یادگنجایی", date=2014-10-12].

To sum up, the form **хофиза** was borrowed from AR to designate 'memory' in general. Then it developed a specific meaning of 'computer memory', possibly under the influence of FA.

---

[53] As پرونده may be used in a sense not related to computers, and فیل may be read /fil/ (meaning 'an elephant'), the word رایانه 'computer' has been added in the searches to obtain more reliable results in the sense of proportions.

Interestingly, **хофиза** in the sense of 'computer memory' may be used in plural: **хофизахо** /hofiza-ho/, Thus, gaining the meaning of 'computer memory chip-sets/devices', like in the compound **USB-хофизахо** [Nokia 2010-07-19: 41].

Although, as it has been said, the form **хофиза** is of AR origin, its proto-source حافظة is not the most frequent form for 'computer memory' in the AR language itself. It is attested [Google search: keyword="حافظة الكمبيوتر", date=2011-09-24] but it is quite rare, if compared to ذاكرة الكمبيوتر [Google search: keyword="ذاكرة الكمبيوتر", date=2011-09-24].

One more word of similar nature (i.e. an Arabism applied to the new idea of 'computer memory') is used in TJ: **хотира** [Komilov & Šarapov 2003: 26ff.] /xotira/.

2.19 Computer mouse

Computer mouse was invented in 1963 by D. C. Engelbart [Yost, 2005, 147]*, who also named the device after the little rodent* [Khazan 2013].

The TJ form **муш** /muš/ is a word used for both animal and the computer control device [Komilov & Šarapov 2005: 29; Komilov & Šarapov 2003: 47; Central Asian Voices acc. 2011-03-20]. It is a calque of a wide spread internationalism based either directly on the EN **mouse** or, which is more probable, RU **мышь**.

The form **мушвора** /muš-vora/ is poorly attested, nevertheless it may be found in some modern electronic sources [Yak Darveš 2008-05-20; Nuraliyon 2011-06--08]. It is derived from the native word for 'mouse' (an animal) but it is enriched with the suffix **-вора** /-vora/. This form exists (and it is far better attested) in the FA lexicographical sources (see below). In fact, one may put forward the hypothesis that this form was borrowed from FA, like e.g. **роёна** (q.v.).

Another attested name for 'computer mouse' in TJ is **мушак** [Warior 2012-11--02; Čarx-i gardun 2012-11-15] /mušak/. It is often used in an izofat phrase with an adjective **компютерӣ** or n. **роёна** as the modifier, as **мушак** alone normally stands either for 'a rocket, missile' (cf. FA موشک), or for 'a muscle'. It may also be used as a diminutive of **муш** [Saymiddinov et al. 2006: 382].

In FA we find a number of forms like ماوس [Haghshenas et al. 2002: 1098; Gacek 2007: 22], موس [Haghshenas et al. 2002: 1098; Gacek 2007: 22], موشی [Gacek 2007: 22], موشوار [Gacek 2007: 22], موشواره [Google search: keyword=موشواره, date=2014-10-12], etc. Most of them are obviously loanwords (ماوس and موس), even if common Indo-European heritage makes them similar to the native word for 'a mouse' (in the sense of an animal), i.e. موش.

2.20 Digital still camera

Even if there were earlier experiments with digital photography, the first truly digital still camera was the Fujix DS-1P by Fujifilm which was presented publicly in 1988 [Tarrant, 2007, 316].

The phrase **дастгоҳи рақамии суратгирй** [Kumita-i Andozi 2011-04-04; Viloyat-i Suḡd 2011-03-30] /dastgoh-i raqami-i suratgiri/ seems to be the most popular TJ equivalent for the EN 'digital still camera'. It has a form of a typical izofat chain. One of its elements, the adjective **рақамй** 'digital' is of AR origin, and it appears in the AR name for the digital [still] camera: الكاميرا الرقمية [Sony acc. 2011-10-02].

The TJ form **аксбардораки рақамй** /aksbardorak-i raqami/ is an izofat phrase consisting of the noun **аксбардорак** /aks-bardorak/ '[still] camera' (the described element) and the adj. **рақамй** 'digital' (the describing element). It is not a very popular form [Google search, keyword=аксбардораки рақамй, date=2014-08-28]. The noun **аксбардорак** is a form of a pretty complex structure. The last process leading to its appearance was certainly affixation, as we find the suffix **-ак** /-ak/, which is often used to form instrumentatives [Perry, 2005, 419]. What remains after rejecting the suffix is a compound consisting of the noun **акс** /aks/ 'a reflection, a photography' and the present verbal stem **бардор** /bardor/ of the verb **бар-доштан** /bar-doštan/ (See Fig. 10 below). The later belongs to the category of the prefixed verbs, so the prefix **бар-** /bar-/ may finally be indicated here. This means that the form in question is associated with the verbal expression **акс бардоштан** 'to take pictures'.

We have to note that this form has to be classified as a rare one, even if it appears in a considerable number of websites, as they are all secondary and contain copies of the same text.

Another form used is **суратгираки рақамй** [Agroinform.tj 2010] /suratgirak-i raqami/. Its structure closely resembles that of **аксбардораки рақамй**. The only difference lies in the head of the phrase. A compound derived from the expression **сурат гирифтан** /surat giriftan/, instead of **акс бардоштан** is used here.

Another variant used in TJ is **суратгираки дичитол**. This one, however, seems to be very rare, as the author of the present work has been able to find one sole source using it [Soros.tj 2010-09-03]. Its word-formational structure is parallel to that of **аксбардораки рақамй**. However, the form **дичитол** it comprises deserves some attention. Its immediate source might have been the FA ديجيتال /diǰitāl/, especially taking into consideration the vowel correspondence of these. In FA we have two rival adjectives: this very same ديجيتال and ديژيتال /diǰitāl/. The latter seems to be an older one and it is obviously taken from FR **digital**, which is proved by the spirant in the second syllable. The original source of both FR and EN forms must have been the LA **digitālis**, however, the meaning 'numerical' was assigned to the FR form under the influence of the EN one. The LA **digitālis** is an adjective derived from the noun **digitus** 'a finger, a toe' [Stevenson 2010: 489].

In FA, a number of different terms is used, e.g. دوربین عکاسی دیجیتال or simply دوربین دیجیتال [Google search: keywords= دوربین دیجیتال, date= 2014-05-10].

2.21 Electric guitar

The first electric guitar was produced by the Rickenbaker Company in 1931 [Achard 1989: 10].

A TJ form **гитараи электрикӣ** /gitara-i elektriki/ is attested [Abrorova 2009-09-30]. It is an izofat phrase parallel to RU **электрическая гитара** (where a compound **электрогитара** is also used), EN **electric guitar**, etc. The head of the phrase (**гитара**) is a borrowing. The modifier (**электрикӣ**) is an adjective of foreign origin, adapted to fit the TJ morphology by adding the adjectival suffix -**ӣ** /-i/.

In online resources, **гитараи электрикӣ** is attested only in a number of inter-related sites [Google search: keywords= "гитараи электрикӣ", date=2014-09-12].

The original source of **гитара** is the GR κιθάρα, κίθαρις, which was borrowed to LA as **cithara** and then, either through DE **Gitarre** or PL **gitara** was introduced into RU [Vasmer 1987: 1,408]. As the TJ and RU forms are identical, RU seems to be the immediate source of the TJ **гитара**. As far as the etymology of the modifier is concerned, see **стансияи электрикии атомӣ**.

Similar forms are used in FA, گیتار الکتریکی [Google search: keyword=گیتار, الکتریکی, date=2011-09-16] and گیتار الکتریک [Google search: keyword= الکتریک, date=2011-09-16].

If we compare the TJ form to the ones found in AR texts, we see one important difference: AR has the word for 'electric' other than the internationalism used in many languages. i.e. کهربائی and uses it forms like the one in question.[54] Hence, depending on the way of adaptation of the borrowed name for 'guitar', we have forms like: قیثارة کهربائیة [Raseloued [2009-03-25]], غیتار کهربائی [DP-News 2010-10-29], جیتار کهربائی [Al-Arabiyya 2011-06-09] etc.

2.22 Electric refrigerator

Electric refrigerators were invented in 1913 [Moan & Smith 2007: 129].

The first TJ form used to designate 'an electric refrigerator' is the word **яхдон** / yax-don/. It may have four different meanings: 1. 'refrigerator', 2. 'dugout'; 3. '(historically) a leather box used to preserve or transport ice'; 4. '(conversational) big wooden chest' [Saymiddinov et al. 2006: 772]. Even though some lexicographical sources do not mention it as a TJ name for 'an electric refrigerator' [Bertel's et al.

---

[54] By the way, the form کهربائی is an old borrowing from Persian: کهربا 'amber (lit. the one that steals straw)' → کهربایی '(made of) amber; magnetic' [Rubinčik 1970: 2,375].

1954: 473; Kalontarov 2008: 319], its use in this sense is attested in various sources [Lutfulloyev et al. 2007: 81, etc.], especially in online publications [Tojnews 2013-07-08; Faraj.tj 2014-05-29]. It seems that both in TJ and in FA this word had been used in the language before actually the electric refrigerators were introduced, of course only with the meanings 2-4. With the introduction of electric refrigerators, new idea has influenced the word and changed its meaning (see *borrowed meaning* [Bussmann 1998: 138]).

In both DA and PŠ the related word يخدان /yaxdān/ is used referring to electrical refrigerator, too [Lebedev et al. 1989: 725], beside يخچال (q.v.).

From the point of view of etymology and word formation, the form **яхдон** is easily analysable. What we have here is the noun **ях** /yax/ „ice" with the suffix **-дон** /-don/ used to create names of various containers (cf. **намакдон, атрдон** etc.). Thus, with the etymological meaning 'an ice container', it is a good parallel to the Eng. **icebox**.

The plural form **яхдонхо** is attested [MMPI 2012-01-25: 68]. Further derivatives based on **яхдон** exist, too, e.g. **яхдондор** 'possessing a refrigerator; refrigerated' [Saymiddinov et al. 2006: 772] /yaxdon-dor/ (a compound comprising also the present stem of the verb **доштан** 'to have'). This form appears in phrases like **вагони яхдондор** 'refrigerated railroad car' [Saymiddinov et al. 2006: 722] / vagon-i yaxdondor/.

Another form derived from **яхдон** is the adj. **яхдонӣ** [Nazarova 2011: 256] / yaxdoni/, used in phrases like **тачхизоти яхдонӣ** [Ibid.] /taǯhizot-i yaxdoni/ 'refrigerating equipment'.

Moreover, the noun **яхдонсоз** 'producer of refrigerators' /yaxdon-soz/ has been found in a comment posted by certain Салим to an article published on the Радиои Озодӣ website [Radyo-i Ozodi 2005-05-11]. A further derivative, an adjective **яхдонсозӣ** [Istad 2012-02-25; Saidakbar 2012-01-19] /yaxdonsoz-i/ is also attested, used in the izofat phrase **заводи яхдонсозӣ** [Ibid.] /zavod-i yaxdonsozi/ 'a factory producing refrigerators' (See Fig. 11 below).

A form semantically similar to **яхдон** (though based on native elements) is to be found in TR: buz dolabı [Alkım et al. 1996: 205].

Another TJ name for 'refrigerator' is the form **яхчол** /yax-čol/. It is a subordinate compound consisting of the nouns **ях** /yax/ and **чол** /čol/ 'a pit, hole'. According to lexicographical sources, it has got the following meanings: 'an underground cold storage room; a dugout [cooled] with ice' [Saymiddinov et al. 2006: 772]. Thus, theoretically, it should not be used while referring to 'electric refrigerator'. However, analysis of the Tajik-language web-sites proves that it is sometimes used so, e.g.:

худуди 120 хонаводаи маскун дар хобгоҳи пешини корхонаи тавлиди яхчоли шаҳри Душанбе бояд то як ҳафтаи дигар манозили худро тарк кунанд. [BBC Persian 2010-12-01].

(About 120 families inhabiting the **хобгоҳ** *(a collective settlement for the workers of a particular enterprise) of the Dushanbe Refrigerator Factory have to abandon their homes until the next week).*

Wherever **яхчол** is used in this sense, it presents an example of the borrowed meaning phenomenon [Bussmann, 1998, 138], cf. **яхдон** above.

However, it is quite striking that – at least in the Internet – it is the BBC FARSI website that seems to be the main user of the word with this meaning. One could suspect a FA influence here, as in FA, where both the lexemes discussed are to be found, it is یخچال /yaxčāl/ (esp. یخچال برقی) that is normally used with the meaning 're-frigerator' [Āryānpur 1375: 1430], while یخدان /yaxdān/ is rather employed to denote a 'a dugout, a kind of a chest, a clay container for water with ice' [Rubinčik 1970: 2,742; Cf. Āryānpur & Āryānpur 1375: 1430]. Moreover, the news in Tajik form only a subsection of BBC Persian and some impact of FA on the TJ subsection is sometimes clearly visible.[55] Thus, we may try to describe the evolution of meanings of the forms in question in the following way: both **яхдон** and **яхчол** belong to the common FA & TJ lexical heritage. They are both used to denote various containers and structures used to preserve ice and/or to store cooled food and water used tra-ditionally in the pre-industrial era. By the time electric refrigerators appeared (early 20th century) the both forms of Persian had been isolated from each other and so the users of TJ and FA made different choices. Now, as the isolation is no longer in place, we note a tendency to introduce unification in this aspect.

Also in Afghanistan, both in DA [Ostrovskiy 1987: 379; Wahab 2004: 171; Leb-edev et. al. 1989: 725] and PŠ [Lebedev et. al. 1989: 725] یخچال is used for refrig-erator. Some non-Iranian languages have borrowed it, too, e.g. AZ /yaxcal/, /yaxjal/ [Householder & Lotfi 1965: 273].

The next form, **яххона** /yax-xona/, is word-formationally similar to **яхчол**. It is a subordinate compound of **ях** /yax/ (the modifier) and the noun **хона** /xona/ 'house' (the head). It is used with the following meanings: '[a] a dugout, a place for storing ice; [b] a refrigerator' [Saymiddinov et al. 2006: 772]. Its use in the sense of 'an elec-tric refrigerator' is attested also by other lexicographical works [Bertel's et al. 1954: 473]. On the other hand, it is hardly present in the online content [Google search: keyword= яххона, date=2014-04-04], which may indicate that – at least nowadays – it is quite rare. Some modern lexicographical sources explicitly state that this form is used in the sense [a] [Kalontarov 2008: 319].

Another TJ word for 'refrigerator', **холодилник** /xolodilnik/, has not been found in the lexicographical works available, however, it seems to be used to cer-tain extant in the everyday language [Satskaya & Jamshedov 2000: 200]. A good example is a comment posted on 2009-05-16 by certain Хайрандеш from Du-shanbe to an article published on the Радиои Озодӣ website [Radyo-i Ozodi 2009--05-06]. As this is a readers reaction, it reflects a more natural language showing no puristic tendencies:

---

[55] See e.g. the use of the word **роёна** (← FA رایانه /rāyāne/) [BBC Persian 2004-05-17], where it had to be glossed as вожаи шиктаре барои «компютер» 'a more chic word for »comput-er«', as it was hardly used by then. Nowadays, the form in question has become relatively popular.

*... Чахон[56] пеш рафта истодааст, дар давлатхои дигар тамоми тачхизотхои хонаги баркианд аз кулфи дар сар карда то коса шустану дигару дигар, мо чи дорем, **Холодилник**, онро хам хама надоранд ...*

*(... The world is making progress, in other countries all the home appliances are powered by electricity (lit. 'are electric') from the lock in the door to (...) etc. And what we have? A **fridge**, and some of us have not got even this ...)*

The plural form of the word, **холодилникхо**, is attested [Google search: keyword=холодилникхо, date=2014-06-12], which is the proof of its grammatical assimilation in TJ.

A related form is attested in UZ: **holodilnik** [Samsung [2009]: 13].

Apart from the ones mentioned above, some more forms are used for refrigerator in FA: ماشین مبرد [Mirzabekyan 1973: 668], سردخانه [Mirzabekyan 1973: 668], سرماگر [Giunašvili 1974: 445].

2.23 Electric shaver

First successful model of an electric shaver (electric razor) that got popularized was that invented by Colonel Jacob Schick, which was patented in 1928 and started to be sold in 1931 [Cole et al. 2003: 143].

The form attested in Tajik is **ришгираки барқӣ** [Salihov 1990: 106] / rišgirak-i barqi/ or **ришгираки электрикӣ** /rišgirak-i elektriki/. It is, however, hardly present in online resources.

The base for these is a compound consisting of the noun **риш** /riš/ 'beard' and the present stem **гир-** /gir-/ of the verb **гирифтан** /giriftan/ 'to get, etc.' This compound is accompanied by the suffix **-ак** /-ak/, which is used to form concrete instrumentatives, like **гӯшак** 'telephone receiver' [Perry 2005: 419]. The resulting form **ришгирак** may form a part of an izofat phrase then, with the attribute meaning 'electric'. Here, either the adjective **барқӣ** [Salihov 1990: 106] (n. **барқ** < AR برق + adjectival suffix -ӣ /-i/) or its synonym **электрикӣ** [Kerimova 1995: 125] is used.

The adjective **барқӣ** /barqi/ may be, in fact, redundant (at least nowadays), as the form **ришгирак** (the AR script ریشگیرک) seems to designate an electric device itself [see Nazarzoda et al. 2008: 2,166].

FA & DA seem to prefer other forms. FA has ریش‌تراش برقی [Haghshenas et al. 2002: 1558] (also more elaborate ماشین ریش‌تراش برقی /māšīn-e rīš-tarāš-e barqī/ [Lebedev et al. 1989: 78]), تیغ خودتراش برقی [Klevcova 1982: 49] or simply خودتراش برقی [Google search: keyword=خودتراش برقی, date=2010-12-04]. In FA we also find ماشین ریش‌تراش [Haghshenas et al. 2002: 1558] and in DA: ماشین ریش [Ostrovskiy 1987: 46].

As far as the other languages of the region are concerned, we find PŠ د ږیری خرپلو برقی ماشین /də žíri xrəyálo barqí māšín/ [Lebedev 1961: 60] (or simply

---

56 Original orthography of the post (with no TJ diacritics) has been retained here.

ماشین /də žíri barqí māšín/ [Lebedev et al. 1989: 78]), UZ **электр устара** [Koščanov et al. 1983: 86].

2.24 Electric vacuum cleaner

The first electric vacuum cleaner is ascribed to J. Spangler (USA) and dated 1907 [Ziółkowska 1997: 101]. *There were earlier models of vacuum cleaners, however, non-electric.*

The two attested TJ forms, i.e. **чангкашак** چنگکشک [Bertel's et al. 1954: 99; Saymiddinov et al. 2006: 699; Nazarzoda et al. 2008: 2,530; Tojnews 2013-07-08] /čang-kašak/ and **гардкашак** گردکشک [Nazarzoda et al. 2008: 1,308; Satskaya & Jamshedov 2000: 216] /gard-kašak/ are very similar both in their word-formational and semantic structure. They are both compounds with a noun meaning 'dust' (**чанг** /čang/, **гард** /gard/) as the first constituent (the modifier) and the present stem **каш-** /kaš-/ of the verb **кашидан** /kašidan/ 'to pull' as the second one (the head) additionally equipped with the suffix **-ак** /-ak/, used to form names of instruments [Perry 2005: 419] (See Fig. 12 below). Apart from lexicographical works, both forms are attested in the Internet resources, however, **чангкашак** is significantly more popular [Google search: keyword=чангкашак, date=2014-04-04; Google search: keyword=гардкашак, date=2014-04-04]. The forms **чангкашак** and **гардкашак** may be potentially calques of the RU form **пылесос**.

Of the two, **чангкашак** is better attested online, while **гардкашак** is to be found only in the Internet re-editions of various lexicographical works [Google search: keyword=гардкашак, date=2014-09-12].

A lengthy descriptive TJ equivalent: **мошинаи гарду чанг чамъ кунанда** is also attested [Satskaya & Jamshedov 2000: 19]. The head of this izofat phrase, the n. **мошина**, is derived from GR (Doric **μᾱχανά** / Attic **μηχανή**), whence it came into LA (**māchina**). From LA comes the FR **machine**, which was borrowed into DE as **Maschine** and the latter is the immediate source of the RU **машина**. What is particularly interesting in the case of this word, is the fact that being at first borrowed in the form of **машина** [Gacek 2014: 156-157]. The form **мошина**, which is clearly ousting **машина** in the most modern sources came into being as a result of the influence of the related FA form ماشین. Interestingly, both the TJ and FA forms show semantic influence of RU in that – statistically – their first meaning is not 'machine' in general, but mostly 'a car (an automobile)' [Gacek 2014: 157].

In DA & FA the dominating form seems to be جاروی برقی [Klevcova 1982: 557; Asadullaev & Peysikov 1965: 746; Ostrovskiy 1987: 299; Fishstein & Ghaznawi 1979: 239]. A variant of that is جاروب برقی [Mirzabekyan 1973: 465] and a similar form: جاروبرقی is also used [Haghshenas et al. 2002: 1881], which is not – like جاروی برقی – a stable syntactical group, but a regular subordinate compound.

In UZ we find the form **changyutgich** [Balci et al. 2004: 362; Samsung [2009], 13]. The Pashto form is دورپاک [Lebedev 1961: 671] or دوڑپاکی /duṛpākay/ [Lebedev et al. 1989: 592].

2.25 Electronic calculator

Their is a discrepancy between various sources, as far as the year of construction of the first electronic calculator is concerned. Nevertheless, we may quite safely assume it happened in the 1940s and one of the earliest electronic calculators was the device constructed by J. V. Atanasoff in 1942 [Crawdford 1988: 146]. The history of tools helping with arithmetic calculations goes back to the antiquity (note e.g. the famous abacus). However, nowadays the only practically used counting devices are electronic calculators (apart from computers differing from them in a number of aspects), so practically the idea of calculator is limited to electronic calculators. And this is to them that the present entry is dedicated (and not – for instance – to their mechanical predecessors).

A compound **хисобкор** حسابکار [Saymiddinov et al. 2006: 690; Nazarzoda et al. 2008: 2,507] /hisobkor/ is used as the TJ name for calculator. This form is quite well attested in lexicographical works, however, it is almost absent from modern electronic publications. It is a derivative based on the noun **хисоб** /hisob/ (← AR حِسَاب), to which the suffix **-кор/-гор** /-kor/-gor/ is added [cf. Perry, 2005, 420]. This suffix is used mainly to produce **agent** nouns, however, incidentally it may also be used to create name of various objects, e.g. **ёдгор** [ibid.]. The AR noun حِسَاب is derived form the radix <ḤSB>, which is also the source of the AR name for 'calculator': آلة حاسبة [Ba'albaki 1999: 144].

In DA a compound containing the same element is used for 'a calculator', namely: ماشين حساب /māšin-hesāb/ [Awde et al. 2002: 76].

The internationalism **калкулятор** [DMT 2011-[04-22]; Mirzob 2010-12-03] is used in TJ, too. The original source of the word is the Latin word **calculator** designating a person performing arithmetical operations [Tokarski et al. 1980: 332]. However, both its phonetics and orthography show distinctive RU features (esp. the letter **я**) all that being an indication that RU served as the vehicular language in this case. Contrary to **хисобкор**, it is difficult to find this form in TJ dictionaries (apart from Bertel's et al. 1954: 177, though it is not clear for the author of the present work, whether 'an electronic calculator' is actually meant there). On the other hand, it is well attested in electronic sources, and its plural form (**калкуляторхо**) may be found, too [MMPI 2012-01-25].

In FA, the forms ماشين حساب [Haghshenas et al. 2002: 182] and حسابكن [Mirzabekyan 1973: 223] are attested.

2.26 Electron microscope

E. Ruska built the first electron microscope in 1933 [Burgess et al. 1990: 189].

Two TJ forms have been found only in a very limited number of sources: **микроскопи электронӣ** [Mažidov & Nozimov 2006, 131; Abdurrahmon acc. 2011-04-04; Čarx-i Gardun. 2011-08-04] /mikroskop-i elektroni/, **заррабини электронӣ** [Jumhuriyat acc. 2011-04-04] /zarrabin-i elektroni/. They both have a parallel structure of an izofat phrase with the adj. **электронӣ** /elektroni/ as the modifier. This adj. is a suffixal derivative of the internationalism **электрон** /elektron/ 'electron' (See **бомбаи атомӣ**, p. 46). The difference between them lies in the n. meaning 'microscope' appearing as the head: it is an internationalism (probably borrowed from RU) **микроскоп** /mikroskop/ in the first instance and a native subordinate compound, **заррабин** /zarra-bin/ (a subordinate compound of n. **зарра** /zarra/ 'a particle, an atom' and the PrsS **бин-** /bin-/ of **дидан** /didan/ 'to see' (the head) – see Fig. 13).

The immediate source of the word **микроскоп** is – most probably – the RU **микроско́п**. The original source is the neoclassical compound **microscope** coined in FR of the GR **μικρός** 'little' and **σκοπ(έω)** 'to watch'. The PL **mikroskop** might have been the vehicular language between FR and RU [Vasmer 1987: 2,621; Groves 1834: 519]. Other sources seem to suggest a different route: FR → DE **Mikroskop** → RU [Černyx 1999: 1,530], which may be supported by the stress on the ultimate syllable in RU.

The entire term **electron microscope**, or – in fact – its DE source form **Elektronenmikroskop** was used for the first time by Knoll and Ruska in their article published in 1932 [Williams & Carter 2009: 4; Knoll & Ruska 1932: 318ff.].

In both cases, the TJ forms for 'electron microscope' seem to be calques of the internationalism used in many languages, in particular in RU: электронный микроскоп. Cf UZ электрон микроскоп [Koščanov et al. 1983: 557]. See also FA میکروسکوپ الکترونی [Haghshenas et al. 2002: 482; Argāni 1364: 179--180] and میکروسکوپ الکترونیک [Mirzabekyan 1973: 322] etc.

2.27 E-mail

It is quite commonly accepted that the first email message was sent on the ARPANET network in 1971 [Poole et al. 2005: 206].

The EN form **E-mail** (also **email**) in its original form (i.e. in the Latin script) is used in TJ [Gacek 2007: 19]. The very same form in the Latin script, i.e. **email** is used in FA, too [Gacek 2007: 19]. Apart from that, variants of this form transliterated into the Perso-Arabic script are attested in FA, like ایمیل and ای میل [Gacek 2007: 19]. In DA the same form is used, pronounced /imēl/ [Awde et al. 2002: 42].

Originally, **e-mail** (or **email**) is a partial abbreviation of the EN phrase **electronic mail**. The component **mail** was borrowed into EN from FR (whence it may have come from Germanic languages), however, it is in EN, where the association of the word with 'post' was developed [Skeat 1993: 265].

As far as the way of **e-mail** into TJ is concerned, we may consider two possibilities, either borrowing directly from EN or via RU, the latter being much more probable than the former.

The plural of this form, **email-xo**, is seldom used in TJ, nevertheless, it is attested [Google search, keyword=email-xo, date=2014-06-12].

Taking into consideration the lack of graphemical assimilation, we shall classify these forms as foreign words in TJ.

Another form used in TJ is **почтаи электронӣ** [Sobirov 2007: 150; Komilov & Šarapov 2003: 71; BRT acc. 2012-01-08] /počta-i elektroni/. It is a stable izofat phrase, where the described element is the RU loanword **почта** 'post'. The original source of this form is the LA **posita (mansio)** (< **positus**), whence it came into IT (**posta**) and from there to PL as **poszta**, and later **poczta** [Fasmer 1987: 3.348].

The describing element is the form **электронӣ** /elektron-i/ 'electronic' [Saymiddinov et al. 2006: 758], which a derivative with the adj. suffix /-i/ added to the internationalism **электрон**.

The original source of the word **электрон** is the GR **ἤλεκτρον** 'amber; an [amber-coloured] alloy of gold and silver' [Černyx 1999: 2,446]. The latter is associated with **ἠλέκτορ** 'shining light; the Sun; the element of fire; [cosmic] fire', itself of non-GR (possibly Middle Asian) origin [Ibid.]. The word was first used in its modern scientific sense in EN by [George Johnstone] Stoney in the last decade of the 19th c. and soon after it was absorbed into FR and DE [Ibid.] It appears in the RU *Брокгауз-Ефрон* encyclopedia in 1904 [Ibid.]. The author of this work believes that the term might have come into RU via FR, as this would explain the stress on the ultimate syllable in the RU form (the prosodic stress of FR may be rendered in this way in RU, while DE and EN have their stress on other syllables). It seems probable that it is via RU that the word **электрон** was borrowed into TJ, though the author is unable at present to put forward phonetical or historical proofs for that.

Another form used in TJ seems is **пости электронӣ** [Кимиёи саодат 2010-06--05] /post-i elektroni/. It is different from the previously discussed phrase in using the form **пост** instead of **почта** and this may be an influence of FA, where the form پست is used [Gacek 2007: 19]. Eventually, this FA word is to be derived, of course, from the very same LA **posita**, however, the VL are different: LA → IT **poste** → FR **poste** → FA.

One more form – closely related to **почтаи электронӣ** – is **почтаи электроникӣ** [Rozi (Šaripov) 2011-10-19] /počta-i elektroniki/. It is definitely less popular than **почтаи электронӣ**. As far as its structure is concerned, the only difference is use of the adj. **электроникӣ** /elektroniki/ in place of **электронӣ** /elektron-i/, which seems to be an imitation of the FA pattern, c.f. پست الكترونيكى (attested beside پست الكترونيک) [Gacek 2007: 19].

2.28 Flamethrower

Although fire has been used for centuries as a weapon, the idea of a modern, man-held flame-thrower is quite new. Its concept was presented to the German high command in 1901. A prototype was tested in 1908 [Tucker & Roberts, 2006, 677].

TJ forms **огнемёт** [Eršov et al. 1942: 108; Osimi & Arzumanov 1985: 612] / ognyemyot/ and **оташпош** [Eršov et al. 1942: 108] /otaš-poš/ have been attested in lexicographical sources since the the first half of the 20th century. **Огнемёт** is doubtless a foreign word or a loanword taken from RU, while **оташпош** /otaš-poš/ is a native subordinate compound, consisting of the noun **оташ** /otaš/ 'fire' and the present stem **пош-** /poš-/ of the verb **пошидан** /pošidan/ 'to throw'. Hence, the whole compound has the etymological meaning 'the one that throws the fire'. Another attested form, **оташандоз** آتشانداز [Osimi & Arzumanov 1985: 612; Say-middinov et al. 2006: 442; Nazarzoda et al. 2008: 2,39] /otaš-andoz/, is structur-ally similar to **оташпош**, the only difference between them being the use of the synonymic verbal stem **андоз-** /andoz/ ← **андохтан** /andoxtan/ 'to throw'. Of these forms only **оташандоз** has been found online, other being unattested in this type of sources [Google search, keyword: **огнемёт**, domain=.tj, date=2010-12-10 & 2014--04-28; Google search, keyword= огнемётхо, date=2014-04-28; Google search: keyword=оташпош, date=2010-12-10 & 2014-04-28].

Another form used is **оташпошанда** [Bertel's et al. 1954: 290] /otaš-pošanda/. As verbal stems in TJ compounds possess participial meaning, this form is – in fact – very close to the already discussed form **оташпош**, however, the verbal stem is substituted here with explicit participle.

Taking into consideration the similarity of the structure of the form **огнемёт** (despite the language difference) on one side, and **оташпош**/**оташандоз**, the latter may be possibly classified as calques of the RU form.

In FA we find parallel forms آتشانداز /ātaš-andāz/ [Asadullaev & Peysikov 1965: 509] and آتشافکن [Haghshenas et al. 2002: 588; Mirzabekyan 1973: 361]. A similar compound, however, with an Arabism as one of its constituents, is used in FA, too: شعلهافکن [Haghshenas et al. 2002: 588]. Another word for 'flamethrower' known in FA is آتشفشان. An etymologically parallel form is attested in TJ (**оташфишон** /otaš-fišon/, also **оташафшон** /otaš-afšon/), as well, however it seems either to possess only adjectival meaning [Saymiddinov et al. 2006: 443; Bertel's et al. 1954: 290] or to designate 'a volcano' [Bertel's et al. 1954: 290]. It is worth noting that the phrase **яроқи оташфишон** (lit. 'fire-throwing weapon') is used in TJ, however, it has the general meaning of 'fire arm' (E.g. *яроқи оташфишон тамғаи "Калашников"* [Миллат, 2008-11-08]).

In DA a specific form is attested, namely الوانداز /alaw'andāz/ [Ostrovskiy 1987: 220; Google search: keyword=الوانداز, date=2010-12-10]. It is a compound structur-ally identical to آتشانداز, but it uses a different word as its first constituent (الو /alaw/ 'flame').

The form **огнемёт** is used in UZ [Koščanov et al. 1983: 711] an KY [Yudahin 1957: 465], as well. Apart from that, Koščanov mentions also the form **ўтсочар** for UZ [Koščanov et al. 1983: 711].

2.29 Floppy disk

Floppy disks were developed by David L. Noble at IBM laboratories between 1967 and 1971 [Roy, 2001, 17/6].

Two similar forms **диски миқнотисии чандирӣ** [Islomov et al. 2005: 4ff.] / disk-i miknotisi-i čandiri/ and **диски магнитии чандирӣ** [Ibid.] /disk-i magniti-i čandiri/ are attested in TJ. They are both izofat chains with the meaning 'magnetic flexible disk', the only difference being the choice between the two synonymical adjectives: **миқнотисӣ** /miknotisi/ and **магнитӣ** /magniti/ 'magnetic'. Wherever the context is clear, the element **чандирӣ** /čandiri/ may be ommitted [Ibid.] A form without the word for 'magnetic' is attested, too, **диски чандирӣ** [DMT 2010-06-02] /disk-i čandiri/.

A loanword from RU **дискета** [Islomov et al. 2005: 9] /disketa/ is attested as well. Its original source is the AmE form **diskette**. The plural form **дискетахо** / disketa-ho/ is used, too [DMT 2010-11-04].

Another form attested is **дискет** [Komilov & Šarapov 2003: 63] /disket/. It may be seen as a variant of **дискета**, as the only difference is the lack of the final /-a/ of RU origin. It may possibly be a recent direct borrowing from AmE. The plural **дискетхо** /disket-ho/ is used as well [Ibid.]

In FA the borrowing دیسک فلاپی (also simply فلاپی) and a hybrid form (partially a borrowing – partially a calque) دیسک نرم are used [Haghshenas et al. 2002: 596].

2.30 Geiger-Müller counter

Geiger and Rutherford constructed first type of their radiation-counting device in 1908. It was capable of counting α-particles and was called 'Rutherford-Geiger' or simply 'Geiger counter'. In 1928 Geiger, working together with Müller, improved the device, making it able to count β-particles, too. This is often called the 'Geiger-Müller counter' [Iliffe, 1984, 42-43]. *In popular use, the difference between Geiger and Geiger-Müller counter is, in fact, neglected.*

A quite complex form **хисобгараки Гейгер-Мюллер** /hisobgarak-i geyger-myuller/ is attested in TJ. However, it has been found in only one original text replicated in numerous electronic publications [Referaty acc. 2011-07-30], which was not available any more three years later [Google search: keyword=хисобгараки Гейгер-Мюллер, date=2014-04-28]. This form is an izofat phrase with the noun

хисобгарак /hisobgarak/ 'a counter' as the head, and a coordinate compound of two proper names **Гейгер-Мюллер** is the modifier. The latter, itself, is a borrowing from other languages, most probably RU, as we see the typical rendering of the German /ʏ/ with /yu/. The word **хисобгарак** /hisobgarak/, in its turn, is a derivative of the noun **хисоб** 'counting, account' with two consecutive suffixes **-гар** /-gar/ (in most cases forming nomina agentis [Perry, 2005, 419]) and **-ак** /-ak/ (often used to create instrumentatives [Perry 2005: 419]). The whole phrase could be possibly classified as a calque of the RU **счётчик Гейгера – Мюллера**.

In FA the forms شمارشگر گایگر [Haghshenas et al. 2002: 652; GRCIR, acc. 2011-07-30; Vahdat11 2010-10-26] /šomārešgar-e gāyger/, شمارشگر گایگرمولر/šomārešgar-e gāyger-muler/ and شمارنده گایگر [Argāni 1364: 229] /šomārande-ye gāyger/ are used [GRCIR, acc. 2011-07-30; Vahdat11 2010-10-26]. Another form attested is آشکارساز گایگر مولر [Tebyan 1392-03-09 HŠ] /āškārsāz-e gāyger-muler/.

2.31 Hair dryer (electric, hand-held ~)

Hand-held, electric hair dryers were introduced in 1920 [Hillman & Gibbs, 1999, 1889].

The form **фен** [Nazarzoda et al. 2008: 1,871; Parvina acc. 2011-08-01] /fen/ is attested in TJ. There is no doubt it is a borrowing from RU, where – in its turn – the form was introduced from German (← enterprise name **Foen**).

Another TJ form used to denote a hair dryer is **мӯхушккунак** موخشککنک [Nazarzoda et al. 2008: 1,871] /mŭ-xušk-kun-ak/. This one is attested in a limited number of online resources, mostly lexicographical websites [Google search: keyword=мӯхушккунак, date=2014-09-12].

A variant of this is form is attested (even if rare): **мӯйхушкунак** [Parvina acc. 2011-08-01] /mŭy-xuš-kun-ak/. The latter is based on a variant form of the noun **мӯ** /mŭ/, i.e. **мӯй** /mŭy/. To analyse these forms, we have to recall the fact that a relatively popular type of compound words (mostly nouns) in all the variants of Persian is constituted by lexicalised sentences (Rubinčik 2001: 164-165), e.g. مرا فراموش مکن 'forget-me-not', or برف‌پاکن 'wind-shield wipers' (lit. 'Clean-the-snow!'). Similar forms are to be found in many languages, EN included (cf. **merry-go-round**), but it is in FA, TJ and DA, where they are quite frequent. A characteristic feature of TJ is that some of these forms take additionally a suffix, in most cases /-ak/, which is the case of **мӯхушккунак/мӯйхушкунак**.

Existence of a similar form in FA (see below) urges us to ask a question whether the TJ form is original or not. One has however to remember that the term in question is a quite obvious descriptive name and it might have appeared independently.

A similar form is used in FA: موخشککن [Google search: keyword=موخشککن, date=2014-10-12] /mu-xošk-kon/, however, let us pay attention to the fact that the instrumentative suffix /-ak/ is not used there. The typical FA equivalent is, however,

سشوار [Haghshenas et al. 2002: 713]. In DA the form /mo-xoškun/ is attested [Awde et al. 2002: 163].

2.32 Hearing aid

First instruments developed to support failing human hearing certainly pre-dated the period covered by this study. However, we shall understand this concept in a narrower sense of electric and portable devices (as these are certainly the features required when we talk about hearing aid today). The first electric hearing aid was patented in 1901 by M. R. Hutchinson, however, it was not portable [Smith 2009: 20]. The first portable (even if not particularly convenient) electric hearing aid was produced by the Marconi company in 1923 [Carlisle 2004: 353].

The TJ form **асбоби шунавой** [Mažlis-i Oli 2010-12-16] /asbob-i šunavoi/ may be probably seen as a calque of RU **слуховой аппарат**. It is an izofat phrase with the noun **асбоб** /asbob/ (formally the pl. of **сабаб** /sabab/, here with the meaning 'instrument, tool') as the head and the abstract noun **шунавой** /šunavoi/ 'hearing' as the modifier. The latter in its turn is a derivative of the participle **шунаво** /šunavo/ created by adding to it the suffix **-й** /-i/ (See Figure 14).

The phrase may be found in a limited number of websites [Muhammadražab 2012-08-01].

A similar form is used in FA دستگاه شنوایی [Asadullaev & Peysikov 1965: 849]. It has, however, rivals, suffixal derivatives سمعک (basing on the AR radix meaning 'to hear', cf. AR المساعد السمعي [Ba'albaki 1999: 418]) [Āryānpur & Āryānpur 1375: 672; Haghshenas et al. 2002: 738] and گوشیار [Āryānpur 1375: 1031].

In other languages of the region we may found periphrastic forms like the UZ **қулоққа тутадиган аппарат** [Koščanov et al. 1984: 445].

2.33 Helicopter

*The history of helicopter construction is in fact quite long and starts about 1500 years ago in China, where a flying toy powered by rotating horizontal wings was invented. Leonardo Da Vinci invented project of his own flying toy based on the same principle. Then, in 1860s Gustave de Ponton d'Amecourt started to build small aircraft models powered by springs. He also coined the word **helicoptère** (← GR 'spiralling wing'). However, true helicopters, as we define them today (an aircraft capable of carrying passengers, powered by combustion engine) were first built in the 20th century. First experimental flight was made in 1907 by Paul Coriv, and the first practical helicopter was constructed by Igor Sikorsky in 1938 [Shaw 2003: 54].*

Чархбол چرخبال /čarx-bol/ is a form attested in lexicographical sources [Nazarzoda et al. 2008: 2,539; Saymiddinov et al. 2006: 702], other written texts [Rahimov et al. 2006: 155] and online [Radyo-i Ozodi 2010-11-10; Google search: keyword=чархбол, date=2011-01-09; Maǯalla-i Bomdodi 2011-03-08, etc.] Word-formationally it is a subordinate compound built of native elements: the n. **чарх** 'a wheel' and another n. – **бол** 'a wing'. From the etymological point of view, however, we may put forward the hypothesis that this is a loanword from the closely related idiom, i.e. FA. This view may be supported by the fact that this form is generally to be found only in the sources from the times after the fall of the Soviet Union.

The parallel form چرخبال is used not only in FA [Google search: keyword=چرخبال date=2015-01-28] but also in DA (An Internet search reveals over 7000 results, some of them from .ir and some from .af domains [Google search: keyword=چرخبال, date=2011-01-09]).

The plural from **чархболхо** is attested in TJ, too [Radyo-i Ozodi 2012-07-17]. The form **чархбол**, together with the PrsS of the verb **бурдан** /burdan/ forms the compound adj. **чархболбар** /čarxbol-bar/ 'carrying helicopter(s)' (like in the phrase **киштии чархболбар** /kišti-i čarxbolbar/) [Radyo-i Ozodi 2010-10-18].

A noun **чархболсозӣ** /čarxbol-sozi/ 'production of helicopters' is attested, too [Sulaymoni 2012-05-11].

The form **вертолёт** ویرتالیات /vyertolyot/ is attested in TJ [Nazarzoda et al. 2008: 1,275; Bashiri, 1994, 127] even if some important lexicographical sources omit this form [Saymiddinov et al. 2006]. The word **вертолёт** does not appear in the military dictionary by Eršov and others (neither does **чархбол**), which was published in 1942. It is not surprising as the total number of helicopters built in the world by that time was not impressive.

Вертолёт is certainly a loanword from RU and it cannot be found neither in FA nor in DA. On the other hand, not surprisingly, it is attested in other languages of the former Soviet Central Asia and Caucasus, e.g. UZ **vertolyot** [Balci et al. 2004: 303], AZ **vertolyot** [Öztopçu 2000: 370], TK **вертолет** [Hamzaev 1962: 125].

The plural form of **вертолёт**, i.e. **вертолётхо** /vyertolyot-ho/ is attested, too [Google search: keyword=вертолётхо, date=2011-03-23].

The internationalism **хуликуптар/хеликуптар** /hulikuptar, helikuptar/ is present only marginally in the lexicography [Nazarzoda et al. 2008: 2,539 – entry **чархбол**] and in the electronic sources [BBC Persian 2009-02-05]. The form is used in FA, as well: هلیکوپتر, هلیکوپتر [Klevcova 1982: 67; Asadullaev & Peysikov 1965: 82] /helikopter/ [Rubinčik 1970: 2,720]. It is also attested in DA هلیکوپتر / hilēkōptar/ [Lebedev et al. 1989: 94], /helikōptar/ [Awde et al. 2002: 42] and in PŠ هلیکوپتر /halikoptár/ [Lebedev et al. 1989: 94]. The form in question goes back to an artificial neoclassical FR compositum **helicoptère**, which was created of two elements of GR origin ἕλιχ, ἕλικος 'a turn, twist, roll, circuit, spiral etc.' and πτερόν 'a wing, a feather' [Spirkin et al. 1980: 117; Groves 1834: 193, 502]. However, the way the form travelled into TJ is difficult to trace. What may certainly be said, the words **хуликуптар** and **хеликуптар** probably have not been transmitted via the RU language. It may be stated, as in RU we would expect the regular change /h/ → /g/. Indeed the form **геликоптер** is attested in RU [Spirkin et al. 1980: 117], even if today it is quite rare. One may speculate that the TJ **хеликуптар** comes from the DA /hilēkōptar, helikōptar/ (see analogies in vocalism). The first vowel of the form **хуликуптар**, in its turn, may be a result of the process of the assimilation (vowel of the first syllable influenced by that of the third one).

The same internationalism (with necessary phonetic alternations) is used e.g. in AR: الهليكوبتر [Ba'albaki 1999: 421].

Some forms unrelated to the above-mentioned are used in FA as well, e.g. بالگرد [Haghshenas et al. 2002: 744].

2.34 Hormone

Although first hormones were discovered in the 19ᵗʰ century, the term has been coined only in 1905 [Brzeziński, 1995, 343].

The EN term **hormone** was first used by Starling in 1905 [Starling 1905: 339ff.] It is derived from the GR **Ορμαω** 'I arouse to activity' [Mutt 1982: 232]. The origin of the RU **гормон** is to be traced back to this EN from. However, it is difficult to indicate a potential vehicular language. We would probably exclude FR with its absence of the initial consonant. Direct borrowing from EN is questionable, as we find no diphthong in the second syllable of the RU form (though this may be due to the nature of scholarly borrowings, often based rather on the written form than the real pronunciation).

The dominating TJ form **гормон** /gormon/ گارمان is attested both in lexicographical works [Bertel's et al. 1954: 104; Osimi & Arzumanov 1985: 184; Saymiddinov et al. 2006: 152; Nazarzoda et al. 2008: 1,330] and in the modern usage [Google search: keyword=гормонхо[57], date=2010-12-11, 2014-05-03]. The phonetics of the form (The initial /g-/ corresponding to the original /h-/) clearly indicates RU as its immediate source. This form is used in other (non-Iranian) languages of the region, like UZ [Koščanov et al. 1983: 215] and KY [Yudahin 1957: 138].

The plural form **гормонхо** /gormon-ho/ is attested, as well [DAT acc. 2011-03-23; Radyo-i Ozodi 2009-10-07]. The form in question may also serve as the base for further derivation, cf. the adj. **гормонӣ** /gormon-i/ [Maǯlis-i Oli 2003-12-08]. Compounds with **гормон** exist, too, like **гормондор** /gormon-dor/ 'containing hormone(s)' (with the PrsS of the verb **доштан** /doštan/ 'to have') [Maǯlis-i Namoyandagon 2001-05-10].

Apart from **гормон**, a related form with no trace of RU pronunciation is attested, too: **хурмун** [BBC Persian 2009-05-14; a comment posted on 2010-01-27 13:23 by certain Далер referring to an article published by Radyo-i Ozodi 2010-01-25] / hurmun/. The plural form **хурмунхо** /hurmun-ho/ is attested, as well [BBC Persian

---

[57]   The sg. form **гормон** is not specific enough. It produces mainly numerous results in RU. Adding the TJ plural suffix -xo helps to focus at the TJ forms.

2009-09-15] and so is the derived adjective **хурмунӣ** /hurmuni/ [Serajtj.com acc. 2011-03-24].

The form **хурмун** may possibly be a borrowing from FA, esp. as it follows the voice rules regarding the correspondences between the FA and TJ vowels.

The FA equivalents of TJ **гормон** are هرمون [Asadullaev & Peysikov 1965: 164] /hormon/, هورمون [Haghshenas et al. 2002: 770] /hormon/ and اورمون [Mirzabekyan 1973: 129]. Apart from that, a native form گیزن is used [Google search: keyword=گیزن, date=2014-10-12]. In DA two forms have been found in lexicographical works: هورمون /hormon/ [Sādiqyār 1379 HŠ: 178; Yussufi 1987: 185; Lebedev et al. 1989: 167] and اورمون [Sādiqyār 1379 HŠ: 133]. Sādiqyār qualifies both forms as taken from FR. This seems understandable in the case of اورمون, where the lack of the initial /h-/ may reflect the FR pronunciation. However, this is not so convincing in the case of هورمون, as there the initial /h-/ is preserved. This may be explained as the influence of the original orthography or, alternatively, this rival form might have been borrowed from EN.

In Pashto we find the form هرمون pronounced /harmón/ [Aslanov 1966: 970] and /hormón/ [Lebedev et al. 1989: 167]. AR has the same internationalism: اَلْهُرْمُونّ [Arslanyan & Šubov 1977: 154; Ba'albaki 1999: 434] beside native حَاثَةٌ [Arslanyan & Šubov 1977: 154].

2.35 Insulin

*The history of the discovery of insulin is quite complicated and so is the history of the name of this hormone. It seems that the name of this hormone has been independently introduced into FR and EN respectively in 1909 by Jean de Meyer (as **insuline**) and later in 1922 (in the form **insulin**) by the Canadian team researching the connection between pancreas and diabetes [Rosenfeld 2002: 2271; Messadié 1995: 112].*

The internationalism **инсулин** /insulin/ is attested in TJ [Maǰlis-i Oli 2009-04-01; Kabirov & Ayubova 2009]. It may be found in numerous electronic publications [Xalili 2011-07-14; Xayrov acc. 2014-07-30; etc.]. Taking into consideration the association of the substance in question with the so called islets of Langerhans, it was quite a natural idea to name this hormone by the name derived from the LA word **insula** 'an island' [Rosenfeld 2002: 2271; see also Davidson 2000: 329; Bilous et al. 2010: 2.7]. Thus, it is not so surprising that the term was introduced by various individuals independently.

Apart from the noun, a derived adjective, **инсулинӣ**, may be found in TJ, too [Google search: keyword=инсулинӣ, date=2014-06-12].

The form **инсулин** is no doubt a borrowing, however, it is hard to indicate possible VL between FR or EN and RU in the case of this form.

In FA, the form انسولین is used [Haghshenas et al. 2002: 834] /ansulin/ beside اینسولین [Mirzabekyan 1973: 212] and in DA انسولین is attested [Sādiqyār 1379 HŠ:

132]. Other languages of the region use a form of the same internationalism, e.g. UZ **инсулин** [Koščanov et al. 1983: 393]. AR has got the loanword الانسولين [Ba'albaki 1999: 472].

2.36 Integrated circuit

The integrated circuit, also refereed to as IC or a chip (a term also used for a micro-processor) was invented in 1958 [Kilby 2000].

Calque of the RU form **интегральная схема** is used in TJ, i.e. **схемаи интегралй** [MMPI 2011-11-16, 20] /sxema-i integrali/. A limited number of oc-curences in online resources may be noted [Google search: keywords= "схемаи интегралй", date=2014-05-20].

Miniaturized integrated circuits may be referred to as **микросхемаи интегралй** [Maǯlis-i Oli 2004b] /mikrosxema-i integrali/. It is worth noticing that the morpheme **микро** (obviously of foreign origin) is productive as a (semi)-prefix in TJ, which is proved by the existence of forms like **микромавч** /mikro-mavǯ/ **микроиқлим** / mikro-iqlim/ [Saymiddinov et al. 2006: 348].

The head of the phrase **схемаи интегралй**, i.e. the noun **схема** 'scheme, cir-cuit' is a borrowing from RU, whereto it came from GR **σχῆμα** via LA **schema** and PL **schema** [Vasmer 1987: 3,815]. The modifier is an adjective, similarly borrowed from RU (**интегральный, -ая, -ое**). The history of this form may be traced back to the LA [Spirkin et al. 1980: 201] **integralis**.

The EN **chip** is borrowed into TJ, too, and it appears there as **чип** (pl. **чипхо**) [MMPI 2011-07-21] /čip/. Moreover, the EN word **chip** is used in TJ in both its original meanings, i.e. not only in the sense of 'integrated circuit', but also 'micro-processor' (q.v.)

There also exists an adjective, comprising the present stem **дор-** of the verb **доштан** 'to have', i.e.: **чипдор** [Abdulloyeva 2010-04-28] /čip-dor/ 'containing an integrated circuit' (e.g. **шиносномаи чипдор** [Ibid.]).

A number of partially related forms for 'integrated circuit' is used in FA: مدار, "مدار مجتمع" and مدار مجتمع also مدار همبسته یکپارچه [Google search: keyword="مدار مجتمع", date=2014-05-12].

2.37 Isotope

F. Soddy discovered existence of types of atoms with different properties in 1913. He also introduced the term 'isotope' in an article written in EN [Encyclopaedia Britan-nica, vol. 6, 420; Baskaran 2011: 3; Soddy 1913: 262ff.].

Soddy's term appears in TJ in the following form: **изотоп** [Saymiddinov et al. 2006: 242; Moukhtor et al. 2003: 84; Habibullayev et al. 2010: 113] ایزاتاپ [Nazarzoda et al. 2008: 1,538] /izotop/. It may be found in online publications [Tojnews 2012-10-12; Vafobek acc. 2014-09-07].

The regular plural **изотопхо** is used, as well [Nazarzoda et al. 2008: 1,538; Mažidov & Nozimov 2006: 162; Habibullayev et al. 2010: 114] /izotop-ho/ and so is the derived adjective **изотопӣ** [Jumhuriyat acc. 2013-01-29].

We do suspect involvement of RU as a vehicular language here, as it is statistically highly probable and as other Central Asian languages of the Soviet Union use similar forms. However, we are not able to show any characteristic phonetic feature proving this. Anyway, EN is excluded as a direct source, because we see no trace of the initial diphthong in TJ. Neither FA seems probable, as the rule FA /ā/ : TJ /o/ is violated here.

The original source of the term is the pseudo-classical EN compound **isotope** artificially coined of GR elements ἴσος 'same, identical' and τόπος 'place' [Černyx 1999: 1,339-340; v. sup.]

In FA, a variant of the same internationalism is used: ایزوتوپ [Klevcova 1982: 240; Asadullaev & Peysikov 1965: 293], though a calque based on native elements: همجای is attested, too [Google search: keyword=همجای, date=2015-01-28] (as for now, not found in TJ sources).

Other languages of the region mostly use the internationalism, e.g. UZ **изотоп** [Koščanov et al. 1983: 383], KY **изотоп** [Yudahin 1957: 250].

2.38 Laser

*Laser (EN **L**ight **A**mplification by **S**timulated **E**mission of **R**adiation) was invented by Th. H. Maiman in 1959* [Brzeziński, 1995, 371].

The form **лазер** لزیر /lazer/ is attested both in lexicographical works [Osimi & Arzumanov 1985: 436; Saymiddinov et al. 2006: 313; Nazarzoda et al. 2008: 1,709] and online [Termcom.tj acc. 2010-07-28]. The original source of this form is the EN acronym 'laser' (see above). However, the form used in TJ resembles the one appearing in RU (**лазер**), so we may suppose that RU served as an intermediary here. This view is further supported by the fact that other languages of the former Soviet Central Asia use the same or similar form, e.g. UZ **лазер** [Koščanov et al. 1983: 504].

The problem of the potential vehicular language between EN and RU is more complicated. The monophtong of the first syllable might have appeared either under the influence of the written form of the EN word or it also may be a trace of some vehicular language, most probably FR.

An adjective derived from the TJ word **лазер** with the suffix **-ӣ** /-i/ exists, namely **лазерӣ** [Osimi & Arzumanov 1985: 436; Myakišev & Buhovsev 2000: 147; Viloyat-i Suḡd 2010-12-10; Radyo-i Ozodi 2006-04-25] /lazer-i/. It is used in phrases like **диски лазерӣ**, **дастгоҳҳои лазерӣ** [Google search: keyword=лазерӣ, domain=.tj, date=2010-12-24].

The FA form is ليزر [Google search: keyword=ليزر, date=2014-11-11] /leyzar/ and the derived adj. ليزرى is attested, too [Haghshenas et al. 2002: 901]. This is obviously based on the EN laser /leɪzər/. The form لازر /lāzer/ [Azizi & Golbān Moqaddam 1346 HŠ: 653] is attested, too. In DA we find ليزر /lēzar/ and لايزر /lāyzar/ [Ostrovskiy 1987: 165]. A different phonetic adaptation of the very same internationalism is to be found in AR: اللازر [Ba'albaki 1999: 514].

2.39 Lie detector

Lie detector (or polygraph) is another invention for which it is difficult to provide one sole date of its first appearance. The view that subtle physiological phenomena may give out a person lying or being guilty of some crime is in fact a very old one. It was widespread e.g. in ancient China. In modern times, polygraphs (psycho-physiological lie detectors recording a number of parameters appeared only in the first quarter of the 20th century. One of its inventors might have been William Moulton Marston. However, his style of work (far from scientific standards) prevents us from indicating them as the inventor, pointing rather at John A. Larson, who presented a working lie detector in 1921 [Segrave 2004: 15-17].

One of the TJ terms for 'lie detector' is **дастгохи ташхиси дурӯғ** [Ayubzod 2011-05-29] /dastgoh-i tašxis-i durŭḡ/. It is an izofat chain – see Fig. 15.

Another attested form, **детектори дурӯғ** [Ayubzod 2011-05-29] /detektor-i durŭḡ/, is a typical izofat phrase. It also seems to be a calque of the RU **детектор лжы**, or even more than a calque, as the head of the phrase is most probably borrowed from RU. It does not seem to be a very popular form, though. An Internet search produced only the article by Аюбзод and some sources secondary to it as results [Google search: keyword=детектори дурӯғ, date=2011-05-31, 2014-05-14].

One more phrase used in TJ with the meaning of 'lie detector' is **дастгохи дурӯғсанч** /dastgoh-i durŭḡ-sanǰ/, attested in the article by Mr. Аюбзод [Ayubzod 2011-05-29] and a number of secondary sources. Technically, it is a much more complicated form than **детектори дурӯғ**. This is an izofat phrase with the native word **дастгох** /dastgoh/ 'apparatus, instrument' as the head and the subordinate compound **дурӯғсанч** /durŭḡ-sanǰ/ as the modifier one. The latter itself is composed of the noun **дурӯғ** /durŭḡ/, which is the modifier and the PrsS **санч-** /sanǰ-/ (← **санчидан** /sanǰidan/) being the head (Fig. 16). A parallel form دستگاه دروغسنج (or simply: دروغسنج) [Haghshenas et al. 2002: 928] is used in FA, beside a similar form: دستگاه دروغیاب (or just دروغیاب) [Haghshenas et al. 2002: 928].

The word **дурӯғсанч** alone may be used as a TJ name for lie detector, as well [Sufi 2012-02-29].

2.40 Microprocessor

The first microprocessor was Intel 4004 developed by the Intel Corporation in 1971 [Reilly, 2003, 139].

The form **микропротсессор** [Komilov & Šarapov 2003: 33; Yunusi 2007: 18] /mikroprotsessor/ is an internationalism showing traces of RU as a vehicular language (the sequence /ts/ is reflecting the RU affricate). It may be found in a number of online resources [Google search: keyword="микропротсессор", date=2014-09--13]. A poorly attested variant **микропратсессор** /mikropratsessor/ [Google search: keyword=микропратсессор, date=2012-04-25] adds another proof to support this, as the /a/ of the third syllable reflects the RU phenomenon of changing the unstressed /o/ into /a/ (*akanye*). This change is not reflected in the RU orthography (cf. **микропроцессор**), so – paradoxically – the TJ orthographical form shows better the actual RU pronunciation. Such a situation is possible, as the rule to write the words taken from RU in accordance with their original (i.e. RU) written form was abandoned in the nineties of the last century [Ido 2005: 5]. Similarly, the Cyrillic letter **ц** was abolished in the 1999, and it is to be rendered either by **с** or by **тс** in the intervocalic position (which is the case here) [Ido 2005: 5].

The RU form itself is – possibly – a borrowing from EN **microprocessor**. Alternatively, one may put forward the hypothesis that DE (**Mikroprozessor**) served as a vehicular language here, as this would account for the affricate in the penultimate syllable.

The plural of **микропротсессор** is attested as well: **микропротсессорхо** [Komilov & Šarapov 2003: 57ff.] /mikroprotsessor-ho/ and an adj. derived from it exists as well, i.e. **микропротсессорӣ** [DMT [2011-12-02]] /mikroprotsessor-i/.

The internationalism in question is used in FA, too: ميکروپروسسور [Haghshenas et al. 2002: 1059].

Another form: **резпардозанда** ريزپردازنده [Nazarzoda et al. 2008: 2,161] /rez-pardozanda/ is used, too, even if it does not seem to be popular at all (an Internet search produces below 5 results [Google search: keyword=резпардозанда, date=2011-07-23]). Word-formationally it is a compound of the adj. **рез** 'small, little' and the pres. part. **пардозанда** /pardozanda/ derived from the verb **пардохтан** / pardoxtan/ (with various meanings, including 'to occupy oneself with sth' [See Saymiddinov et al. 2006: 456]). This form is much better attested in FA as ريزپردازنده [Haghshenas et al. 2002: 1059] (together with a related one: ريزپردازگر [Google search: keyword=ريزپردازگر date=2015-01-29]) and it seems that it was coined there. In other words, the TJ **резпардозанда** seems to be a loanword from FA[58], where it appeared as a calque of the EN **microprocessor**.

---

[58] The opinion of Собиров that this is a native TJ form [Sobirov 2007: 150] does not seem to be well founded.

Also the EN word 'chip' is used in TJ: чип [BMT 2011-05-20, 2] /čip/ in the sense of 'microprocessor'. However, just like in its source language, it may also have the meaning of 'integrated circuit' (q.v.). It might have been borrowed either directly form EN of via RU.

2.41 Microwave oven

In 1946 Percy Le Baron Spencer observed that microwaves induce vibrations of particles inside objects (i.e. raise their temperature). First microwave ovens using this phenomenon were sold in 1947 [Messadié, 1995, 127].

The form **оташдони микромавч** /otašdon-i mikromavǯ/ is attested in lexicographical works [Nazarzoda et al. 2008: 1,800], however, it has not been found in online resources [Google search, keyword="оташдони микромавч", date=2014--08-15]. This is an izofat phrase with the n. **оташдон** (itself derived from the noun **оташ** 'fire' using the suffix -**дон**) 'oven' as the described element and the form **микромавч** as the describing one. The later is derived from the word of AR origin **мавч** /mavǯ/ using the (semi)-prefix (Cf. Präfixoide *anti-* [Rzehak 2001: 357].) **микро** /mikro/ 'micro' borrowed evidently from RU. The primary source of the latter is the GR **μῑκρός** [Černyx 1999: 1,530]. However, indicating its vehicular language(s) is difficult. Like in the case of numerous affixes of GR origin in various modern languages, the morpheme **микро** had been most probably first absorbed into RU as a part of various lexemes (possibly via different intermediary languages) only later to become a productive suffix in TJ. As the oldest RU words containing this morpheme were probably borrowed from DE and FR, we may put forward the hypothesis that one of them served as an intermediary here [cf. Černyx 1999: 1,530].

It is worth noting that the morpheme **микро** is to some extent productive in TJ (cf. The chapter on **Integrated Circuit**).

The form **микромавч** is a calque of the RU **микроволна**, itself a calque of the EN **microwave**. The whole phrase **оташдони микромавч** seems to be a calque, too (← RU ← EN).

The other variant, **печка[и] микромавч** /pečka-[i] mikromavǯ/ has been found in one sole source [Samsung [2009], 13]. It differs from the previously discussed form only in one aspect: the native form for 'oven' (**оташдон**) is replaced by the RU loanword **печка**. Both forms seem to be calques of the RU **микроволновая печь**.

Three more forms are mentioned in the TJ Wikipedia article on microwave oven, **дастгохи майкровейв**, **тундпаз** and **фармавчпаз** [Wikipedia: entry=дастгохи майкровейв, date=2014-08-28]. They are, however, to be found only in derived online resources [Google search: keywords="дастгохи майкровейв", тундпаз, фармавчпаз; date=2014-09-12]. As far as their structure is concerned, **дастгохи майкровейв** /dastgoh-i maykroveyv/ is an izofat phrase with the native noun **дастгох** as its head and the borrowed (from EN via RU) modifier **майкровейв**.

Тундпаз is a compound based on native elements **тунд** 'fast' (adj. & adv.) and **паз-**, the present stem of the verb **пухтан** /puxtan/ 'to cook'. Finally, **фармавчпаз** is a compound of **фармавч** (itself a derivative of **мавч** 'wave' built by addition of the prefix **фар-** 'above') and the above-mentioned verbal stem **паз-**.

The FA equivalents for the EN 'microwave oven' include خوراک‌پز مایکروویو (or simply: مایکروویو) [Barzgar 1390-07-04], اوجاق مایکروویو [Haghshenas et al. 2002: 1059] and a number of less popular forms. In UZ we find **micromavj temir** [Samsung [2009], 13].

2.42 Monitor (= computer display)

It seems that first computer-display systems (and hence monitors) were used in the 1950s as a part of the SAGE air-control system [Manovich 2002: 101].

The TJ word **монитор** [Moukhtor et al. 2003: 144; Saymiddinov et al. 2006: 354; Baizoyev & Hayward 2004: 345; Komilov & Šarapov 2003: 34] /monitor/ is an internationalism and parallel forms are be found in many languages (including FA مانیتور /mānitor/, which has the native نمایشگر /namāyešgar/ as its rival [Haghshenas et al. 2002: 1087; Gacek 2007: 19]). It is attested in TJ online sources [Matrix-tv acc. 2014-09-12; Muhammadsodiq 2014-09-11].

There is also another – very similar – form in TJ, i.e. **манитор** /manitor/ [DMT 2010-04-15]. In this case, however, we notice an interesting phonetic detail – the vowel /a/ in the first syllable. Theoretically, it may be a reflection of the pronunciation of the EN **monitor** (BrEN /mɒnɪtə/, AmEN /mɒːnɪtər/. However, it is much more plausible that it is a rendering of the actual RU pronunciation of the word **монитор**, i.e. /manitór/[59]. We do find orthographic form **манитор** (an Internet search results in over 300000 web pages containing this form [Google search: keyword=манитор, language=Russian, date=2011-03-20].

According to Спиркин, this form was borrowed into RU from EN [Spirkin et al. 1980: 329], though one may suppose there might have been some intermediary between the two idioms. The EN **monitor** is derived from the LA **monitor** 'counsellor, preceptor; prompter'. It seems that the TJ word for 'monitor' (in the sense of computer display) entered the vocabulary in two phases. First the word was borrowed with its original meanings. Than, under the influence of RU (or EN directly) it changed its meaning to comprise the idea of Video Display Unit.

Related forms are used in other languages taken into consideration, cf. AZ/UZ/TK **monitor** [Samsung [2009]: 11, 13], KY **монитор** [Samsung 2009: 13].

---

[59] The change of the unstressed vowel /o/ into /a/ is a typical feature of the leading RU dialects.

2.43 Mp3 player

The first mass-produced MP3 player appeared in 1998 [Betz 2011: 264].

A loanword **МР3-плеер** [Tcell acc. 2012-04-15] / **МП3-плеер** [MMPI 2011--10-24] is attested in TJ. Its original source is, surely, the EN **MP3-player**, however, as the identical form exists in RU (**MP3-плеер**), it is highly probable that the latter served as an intermediary here. The abbreviation forming the first part of the form is written either in Cyrillic or in the Latin script. The first variant should be classified as a foreign word, the latter, possibly, as a loanword.

In other languages of the former Soviet Central Asia and adjacent regions parallel forms are attested, cf. AZ **MP3 pleyer** [Samsung [2009]: 11], KA & KY **MP3-плеер** [Samsung 2009: 13], TK **MP3-pleýer** [Zaman Türkmenistan 2009-06-16; Samsung 2009: 13], UZ **MP3-pleer** [Voy 2009-12-29; Samsung 2009: 13].[60]

The form **MP3-бозигар** [Samsung [2009]: 13] is a partial calque, i.e. the initial element '**MP3**' has been retained in its original EN form, while the second element (**бозигар** /bozigar/) is a calque corresponding either directly to the EN player.

2.44 Neutron

The neutron was discovered in 1932 by J. Chadwick [Encyclopaedia Britannica, vol. 8, 625]. *However, as this type of a particle was theoretically predicted somewhat earlier, the term 'neutron' seems to have appeared for the first time in Harkins's article Natural Systems for the classification of isotopes... in 1921* [Herwig 2009: 3].

The TJ word for 'neutron' is **нейтрон** [Saymiddinov et al. 2006: 402; Nazarzoda et al. 2008: 1,906; Normurod & Qodiri 2005: 278] نیترون [Nazarzoda et al. 2008: 1,906] /neytron/, the immediate source of which is the RU form **нейтрон** (See esp. the dyphtong in the first syllable). The word is well attested in online resources [Radyo-i Ozodi 2007-05-15; Nurob acc. 2014-09-12].

Parallel forms exist in other languages of the region: UZ [Koščanov et al. 1983: 645], KY [Yudahin 1957: 416]. The original form was coined in EN (see above) basing on the LA **neutrum** [Tokarski et al. 1980: 509]. The author of this work believes that the word must have been borrowed into RU via DE **Neutron**, /nɔɪtrɔn/, as this would explain the diphthong in the first syllable.[61] Otherwise one would expect the

---

[60] There seems to be a mistake in the forms provided by the latter of the cited sources [Samsung 2009], i.e. we find here **MR3-pleýerler**, **MR3-pleerlar** instead of the expected **MP3-pleýerler**, **MP3-pleerlar**. This may result from the hesitation between the Latin & Cyrillic script.

[61] In such a context the opposition between the original diphthongs like [aɪ], [eɪ] and [ɔɪ] may be neutralized, c.f. DE Neusilber → RU нейзильбер [Spirkin et al. 1980: 338], EN nail → RU нейл.

original initial syllable /nyu:/ of the EN **neutron** to be preserved in RU, just like in the case of EN **Newton** → RU **ньютон** [Spirkin et al. 1980: 348] or FR **nuance** → RU **нюанс** [ibid.].

The word is evidently assimilated as it is used with the plural ending: **нейтронхо** /neytron-ho/ [Boqizoda acc. 2011-03-25; Habibullayev et al. 2010: 111] and it may serve as the base for suffixal derivation, cf. the adjective **нейтронӣ** /neytron-i/ with the suffix -**ӣ** /-i/ [Safar 2010-11-12] (cf. DA نیوترونی [Lebedev et al. 1989: 375]). More complex forms basing on **нейтрон** exist, too, e.g.: **нейтронборонкунӣ** [Habibullayev et al. 2010: 112; Mažidov & Nozimov 2006: 173] /neytron-boron-kun-i/, 'bombardment [of the nucleus of an atom] with neutrons', the structure of which is a result of both composition and suffixation (see Fig. 18 on p. 128).

In FA a number of related forms is attested: نوترون /notron/ [Asadullaev & Peysikov 1965: 459; Rubinčik 1970: 2,670; Omid 1373 HŠ: 1173; Google search: keyword=نوترون, date=2010-12-13]), نیترون [Google search: keyword=نیترون, date=2010-12-13; Asadullaev & Peysikov 1965: 459], which is not very popular, but nevertheless attested in modern usage[62]. Another form used is نیوترون [Giunašvili 1974: 232; Google search: keyword=نیونرون, lang=FA, date=2010-12-15].

Minor differences between the FA forms indicate that they are in fact rival loanwords from various sources (even if the original, primary source is the same). The form نوترون /notron/ seems to be borrowed from FR (FA /notron/ ← FR /nøtʁɔ̃/). The vowel /o/ of the FA form is the closest parallel to the FR monophtong /ø/[63], the same is true about FA /r/ and FR /ʁ/[64] and the nasalized /ɔ̃/ is rendered by /on/, possibly with some influence of the written form[65]. The form نیوترون /nyutron/ in its turn is most certainly taken from EN 'neutron' /njuːtrɔn/, which is revealed by the identical cluster of the glide /y/ and the vowel /u/.

In DA we find نیوترون [Sādiqyār 1379 HŠ: 173][66]. In AR the form النیوترون is attested [Baʻalbaki 1999: 611].

2.45 Nuclear power station

The first nuclear power station in the world was opened in England in 1957 [Steed 2007: 220].

Probably the most popular form for nuclear power station in TJ is **нерӯгохи атомӣ** [Radyo-i Ozodi [2010-08-13]; Muhammad 2011-01-20] /nerûgoh-i atomi/ (See Fig. 19). The plural form of this, **нерӯгоххои атомӣ**, is attested as well [Ozo-

---

[62] The initial syllable may indicate RU or – possibly – DE as the immediate source of this form.

[63] Cf. FR (← GR) **eucalyptus** /økaliptys/ → FA اوکالیپتوس /okāliptus/.

[64] Cf. FR **allergie** /alɛʁʒi/ → FA آلرژی /ālerži/.

[65] Cf. FR **compteur** /kɔ̃tœʁ/ → FA کنتر /kontor/ [Omid 1373 HŠ: 986].

[66] Sādiqyār marks it as a loanword from FR, see, however, the discussion on the parallel form in FA above.

dagon 2011-06-09]. It also appears in a more elaborate form as **нерӯгохи баркии атомӣ** [Mažidov & Nozimov 2006: 59] /nerûgoh-i barqi-i atomi/.

The word **нерӯгох** is a subordinate compound of two substantives, **нерӯ** 'power' and **гох** 'place' (the head)[67].

The adj. **атомӣ** may be replaced by its native synonym **хастай**, hence **нерӯгохи хастай** [Ayubzod 2011-05-02; Abdullohi 2011-05-12] /nerûgoh-i hastai/. A parallel form exists in FA: نیروگاه هستهای [IRNA 1390-04-02 HŠ; Fardanews 1390-01-02 HŠ]. The **хаставӣ** /hasta-vi/ variant is attested, too: **нерӯгохи хаставӣ** [Qodir 2007-11-29] /nerûgoh-i hastavi/ (mostly in the Radyo-i Ozodi materials and secondary sources referring to them).

As far as the adjectives **атомӣ** and **хастай/хаставӣ** are concerned, see respectively **бомбаи атомӣ** (p. 46) and **бомбаи хастай** (p. 46).

The same form, نیروگاه اتمی, is used in FA [Haghshenas et al. 2002: 1150; Sahām Nyuz 1390-03-24 HŠ] and in DA [Bakhtarnews 2011-04-11; Awde et al. 2002: 206].

Another form: **стансияи электрикии атомӣ** [Habibullayev et al. 2010: 114] /stansiya-i elektriki-i atomi/ is used, too. It is a sequence of izofat phrases, the so called izofat chain. The first element is obviously taken from RU **станция**, especially if we take into consideration the fact that it was previously written in accordance with the original orthography (cf. **стансияи электрик** [Eršov et al. 1942: 179]). The adj. **электрикӣ** /elektrik-i/ is an internationalism borrowed from RU with the adjectival suffix /-i/ added and thus adapted to fit better the TJ lexical corpus (see again the Eršov's example above, where the form **электрик** /elektrik/ is used, with the suffix not added yet).

It is worth noting that the complex phrase **стансияи электрикии атомӣ** should be analysed rather as a sequence of two subsequent izofat phrases: /(stansiya-i elektriki)-i atomi/, the second one providing an attributive modifier for the whole first phrase, rather than a chain of two attributes referring to the base noun. To support this view we may put forward two arguments: it does not seem to be an option to change this chain into a phrase with the conjunction /va/: */stansiyai elektriki-vo atomi/, which should be the case, if /elektriki/ and /atomi/ were two equal level attributes of the noun /stansiya/. Moreover, **стансияи электрикӣ** seems to be an existing stable collocation.

When the context is clear, the element **электрикӣ** /elektriki/ may be omitted, thus, leaving **стансияи атомӣ** [DMT 2010-05-14] /stansiya-i atomi/. An acronym of the form **стансияи электрикии атомӣ**, i.e. **СЭА**, is used in TJ as well [Bobiyev et al. 2007: 214].

As far as the etymology of the elements of this form is concerned, they are all of foreign origin. On the analysis of the adj. **атомӣ** see **бомбаи атомӣ** (p. 46). The form **стансия** is a heterogeneous one. Its immediate source is the RU **станция**,

---

[67] It is worth noting that the status of the element roх is disputable: it may also be classified as a suffix or semi-suffix. For the discussion on this problem in FA see Рубинчик [Rubinčik 2001: 153].

where it is derived from the older native **стан** that has been influenced by western European derived from the FR **station**, i.e. EN **station**, DE **statión**, etc. The FR form in its turn goes back to the LA **stātiō**, Gen. **stātiōnis** 'place, location; state' [Černyx 1999: 2,198].

The origin of the form **электрикӣ** is even more complicated. It is derived by the means of sufixation from **электрик,** which is used in TJ as both n. 'electricity' and adj. 'electric') [Bertel's et al. 1954: 465]. Bashiri states that **электрик** is a borrowing ← RU **электрик**, just like in the case of TJ **химик** 'chemist' ← RU **химик** (with the same meaning) [Bashiri 1994: 124]. The problem is that the difference in the meaning of the RU and TJ **электрик** makes it rather unlikely (the RU form stands for 'electrician' [Kovtun & Petuškov 1965: 1809-1810]). It seems much more probable, in the humble opinion of the author of this work that the TJ form is based on either EN **electric** or FR **électrique**. It may be a scholarly borrowing, as opposed to natural and spontaneous one. The origin of this term may be traced back to the GR **ἤλεκτρον** and **ἠλέκτορ**, via the LA adj. **electricus** derived from the noun **electrum** (see the chapter on **electronic mail**).

We find similar forms in other languages of the former Soviet Central Asia, e.g. UZ **атом электр станцияси** [Koščanov et al. 1983: 43], which is built of exactly the same elements, however, in accordance to UZ syntax.

In DA, on the other hand, we find a form quite similar in its structure, however, with the EN equivalent of the RU **стансия**, i.e. استیشن اتومی /istēšan-e atōmī/ [Lebedev et al. 1989: 52; 664].

The PŠ form is close to the TJ **стансияи электрикии атомӣ**, as far as the lexical elements used are concerned. The syntax of this word group is, however, different and – at the same time – typical for PŠ, د برېښنا اتومی تېسن /də brešnā aṭomí ṭesən/ [Lebedev et al. 1989: 52; 664; 756].

2.46 Nylon

Nylon was first produced in 1938/1939. The name was most probably coined from the initial pars of the toponym New York and London [Ziółkowska 1997: 129], though some scholars contest this etymology [Černyx 1999: 1,597].

The TJ form **нейлон** نیلان /neylon/ is identical with the one used in RU, so it may be assumed with great probability that RU was the immediate source for TJ (EN being the original one), especially when we take into consideration that both TJ and RU forms contain the same diphthong [-ey-], even if the inherited diphthong [-ay-] exists in TJ and that could have been used to reflect better the [aɪ] of the EN original. There was probably one more vehicular language between EN and RU, namely DE [Černyx 1999: 1,567]. Such a change of the original [aɪ] into [ey] may be frequently found in RU loanwords from DE and EN, e.g. DE **Leitmotiv** → U **лейтмотив** [Spirkin et al. 1980: 281], DE **Meistersinger** → **мейстерзингер[ы]** [Spirkin et al. 1980: 309], EN **liner** → RU **лейнер** [Spirkin et al. 1980: 281].

The TJ form **нейлон** appears in lexicographical works [Nazarzoda et al. 2008: 1,906] as well as in online sources [Muhabbat va Oila 2010-07-01]. No plural form of this word has been found, however, an adjective derived from this form: **нейлонӣ** /neylon-i/ [Satskaya & Jamshedov 2000: 107; Nisaun, 2009-03-14].

In FA we find the form نايلون [Asadullaev & Peysikov 1965: 459; Argāni 1364 HŠ: 357] /nāylon/, which clearly reflects the original EN pronuciation /naɪlɒn/. نيلون is attested as well [Mirzabekyan 1973: 345]. In DA we find the form نيلون /naylūn/ [Fishstein & Ghaznawi 1975: 121]. Latify mentions an enigmatic DA form سند (sic!) as an equivalent of the EN 'nylon' [Latify, 1972, 90].

As far as the other languages of the region are concerned, we find PŠ نيلون /naylón/ [Lebedev 1961: 408] and UZ **neylon** [Balci et al. 2004: 175].

2.47 Prion

*Prions were discovered by S. Prusiner in 1982 (hypotheses related to them were present as early as in 1980) who also coined their name from the words **protein** (← FR **protein**[68] ← GR πρωτεῖος 'primary' ← πρώτος 'the first' [cf. Spirkin et al. 1980: 415) and EN **infection** (← FR **infection** ← LA **infectiō** 'a plague' [Černyx 1999: 1,354]) [Messadié 1995: 181-183].*

In fact, the TJ form **прион** /pryon/ is not well attested, nevertheless it has been found in official publications of the Parliament of Tajikistan [Maǧlis-i Oli 2010-12-08] together with its plural form **прионхо**. Apart from the above-mentioned law, which has been published online, the form is difficult to find in world wide web resources.

The same internationalism is used in RU. Moreover, a parallel form is to be found in FA in the form پريون, together with is plural پريونها and derivatives like the adj. پريونى [Qā'edi & Ša'bani 1387 HŠ: 1339-1349]. Taking into consideration this congruence of forms together with the fact that the borrowing belongs to the newest layer, prevents us from assuming automatically that it is RU that played role of the vehicular language. There are other possibilites, including direct borrowing from EN or via FA (It may not be altogether excluded even though the traditional FA /ā/ : TJ /o/ correspondence is violated here).

In numerous languages we find the same internationalism, cf. TR **prion**, pl. **prionlar** [Özdemir 2002]; AZ pl. **prionlar** [ATU acc. 2014-05-15], AR البريون, pl. البرونات (SFDA 2009-04-19).

---

[68] The word protein was introduced by Swedish chemist Jöns Jacob Berzelius in his letter written on July 10, 1838 to Gerrit Jan Mulder [Vickery 1950: 387] and as the correspondence between the two scholars was conducted in FR [Rosenfeld 1982: 6], we assume the term was first used in this language.

2.48 Short Message System (SMS)

The acronym SMS refers both to the system and to the messages sent using it. The first SMS was sent in 1992 and the service became commercially available in 1993 [Safko & Brake 2009: 396].

The EN acronym **SMS** is used in TJ [Ahmadi 2011-06-09; Khovar [2010]-04-09] and is pronounced /esemés/ [Radyo-i Ozodi 2011-12-01]. The form is a very recent one and pronunciation does not indicate the immediate source in a decisive manner. We may put forward two hypotheses. Either the form was borrowed directly from EN, or RU played the role of the vehicular language.

A derived adjective with the suffix /-i/: **SMS-ī** is attested, too [Ahmadi 2011-06-09]. The plural form of SMS, i.e. **SMS-xo** is used in TJ, as well [Ozodagon 2011-07-20; BBC Persian 2009-09-15b].

As the word is not assimilated graphemically in TJ, we shall classify it as a foreign word, rather than as a loanword.

The same EN abbreviation is used in TJ also in a cyrillicized form **смс**, together with plural (**смс-хо**) and the adjective derivative **смс-й** [Tojnews 2013-04-01]. This one will be classified as a borrowing.

Apart from these acronyms, the native form **паёмак** [BBC Persian 2009-09-15b; Karim 2011-01-26] /payomak/ is also used. It can be analysed as **паём** /payom/ message with the diminutive suffix **-ак** added. Plural form **паёмакхо** is attested as well [Růzgor 2012-01-10].

A hybrid form based on both the borrowing **SMS** and the native noun **паём** 'message', i.e. **SMS-паём** exists, too [Čarx-i Gardun 2012-06-08; Orifi 2012-09-07]. The plural form of this, i.e. SMS-**паёмхо** is to be found, as well [Tcell 2011-07-18].

Another attested hybrid form is **SMS-паёмак** [Karim 2011-01-26].

A counterpart of the FA پیام کوناه [Bānk-e Teǧārat acc. 2011-08-09], i.e. **паёми кутох** has been found, too. It appears mostly in a number of electronic publications, most probably translated from FA [IQRA, 2008-12-17]. It seems quite probable that the form has been borrowed from FA.

2.49 Superconductor

Superconductors and the phenomenon of superconductivity were discovered in 1911 [Messadié, 1995, 148].

The TJ form for superconductor is **фавкуннокил** فوق الناقل [Osimi & Arzumanov 1985: 993; Saymiddinov et al. 2006: 621; Nazarzoda et al. 2008: 2,379] /favq-un-noqil/. It is particularly interesting as it is an example of a model quite rare in the analysed corpus, i.e. an AR *'iḍāfa* phrase with a noun in status constructus. The head of the phrase (**фавк** /favq/) is of AR origin and it is used in TJ as a preposition 'above', as a noun 'the upper part', or with the adjectival meaning 'upper' [Saymiddinov et al. 2006: 621]. The modifier is **нокил** /noqil/ 'the one who transports something; conductor', which originated from the AR participle. Between the two, we notice the element **-у** /-u/ which is the AR nominative ending and the AR definite article **н-** /n-/ (← /l-/ ← /al/). There is a number of parallel forms in TJ, like **фавкуттамом** /favq-u-t-tamom/ 'the highest' or **фавкуттассавур** /favq-u-t-tassavur/ 'non-imaginable, beyond imagination'. To sum up, we have two Arabic words here, used in an Arabic syntactical construction (a grammatical borrowing) but still the whole phrase was created in TJ.

The form is hardly present in online resources, apart from some lexicographical sites [Google search: keyword=фавкуннокил, date=2014-08-01].

The plural form of **фавкуннокил**, i.e. **фавкуннокилхо** [Rahimov et al. 2006: 211] is attested, too. The abstract noun ('superconductivity') is **фавкуннокилият** [Rahimov et al. 2006: 210ff.] /(favq-u-noqil)-iyat/ with the assimilated suffix of AR origin.

The AR definite article in the form **фавкуннокил** /favq-un-noqil/ is transcribed with the Cyrillic script according to the actual pronunciation, i.e. it shows the regular (in this context) AR assimilation of /l/ to /n/. However, the form following the AR orthography rather that pronunciation may be found in a TJ publication issued in Uzbekistan, i.e. **фавкулнокил** [Habibullayev et al. 2010: 116].

In FA the forms used to designate a 'superconductor' are ابررسانا [Haghshenas et al. 2002: 1710] and فوقرسانا [Mirzabekyan 1973: 409].

2.50 Tank

The tanks appeared in the battlefields of Europe during the WWI, the first being those constructed by the British, like Little Willie and Mk I. Other countries (esp. France) worked on similar projects at the same time, too. All that happened between 1915 and 1916. The name was coined in 1916 and it was totally arbitrary [Chant 2004: 9, 49].

The internationalism **танк** تنک /tank/ is used in TJ [Eršov et al. 1942: 183; FTAF 1941: 53; Bertel's et al. 1954: 380; Moukhtor et al. 2003: 236; Saymiddinov et al. 2006: 581; Nazarzoda et al. 2008: 2,308; Satskaya et al., 2007, 95].

The word is well attested in the online TJ corpus and the search for this form together with the noun **чанг** /ʒang/ 'war' produces nearly 6000 results [Google search: keywords=танк, чанг; date=2014-05-02].[69]

There is a great number of derived forms based on this loanword, like **танкӣ** تنکی /tank-i/ – an adj. 'of a tank' [Nazarzoda et al. 2008: 2,308]; **танкшикан** /tank-šikan/ 'anti-tank' [Saymiddinov et al. 2006: 581; Nazarzoda et al. 2008: 2,308] – a subor-

---

[69] The orthographic form танк is used in many languages including RU, so another form has to be added as a keyword to ensure that only (or mostly) TJ results are produced.

dinate compound consisting of the noun **танк** (the modifier) and a PstP **шикан-** / šikan-/ (← **шикастан** /šikastan/ 'to break, to destroy, etc.'; the head). The compound **танкшикан** has got a participial meaning of 'tank-destroying, destroying tanks'. Another compound **танкзан** تنکزن /tank-zan/ has a parallel structure, the only difference being the fact that a different verbal stem is used as the head, i.e. **зан-** /zan/ (← **задан** /zadan/ 'to hit'). Semantically, **танкзан** may be a synonym to **танкшикан** or it may have a nominal meaning 'tank attack' [Nazarzoda et al. 2008: 2,308]. Another form – **танкрон** تنکران /tank-ron/ [Saymiddinov et al. 2006: 581; Nazarzoda et al. 2008: 2,308; Bashiri 1994: 130] – has got a parallel structure (a verbal stem used in this case is **рон-** /ron-/ (← **рондан** 'to drive'), however, it is a noun and its meaning is 'tank driver; tankman'. A further derivative based on the latter is attested, namely an abstract n. **танкронӣ** تنکرانی /tankroni/ 'tank-driving' [Saymiddinov et al. 2006: 581; Nazarzoda et al. 2008: 2,308].

Another subordinate compound with the noun **танк** as the modifier and with the structure similar to the ones previously discussed is **танксоз** تنکساز /tank-soz/ 'tank-constructor' [Saymiddinov et al. 2006: 581; Nazarzoda et al. 2008: 2,308] (an abstract derivative with the suffix **-ӣ** /-i/ is known in this case, too → **танксозӣ** تنکسازی /(tank-soz)-i/ 'tank-construction' [Saymiddinov et al. 2006: 581; Nazarzoda et al. 2008: 2,308]).

Another form based on the n. **танк** is **танкчӣ** تنکچی /tank-či/ [Saymiddinov et al. 2006: 581; Nazarzoda et al. 2008: 2,308].[70] It contains a suffix of Turkish (sensu largo) origin productive (at least at some point) in the three idioms in question (DA/FA/TJ).

There are also forms that were clearly borrowed in their entity from RU, like **танкодром** /tankodrom/ [Saymiddinov et al. 2006: 581], where the joining element /-o-/ indicates the immediate source of the form[71]. Some other forms of RU origin were used, too, like **танкист** /tankist/ 'tankman' [Eršov et al. 1942: 183; Bertel's et al. 1954: 380], however, now their rival hybrid forms like **танкрон** seem to be dominating. Nevertheless, the very existence of the forms like **танкодром** or **танкист** hints at RU as the immediate source of the form **танк** itself.

As long as the word **танк** and its derivatives are dominating in lexicographical sources, a similar form – **тонк** /tonk/ (pl.: **тонкхо** /tonk-ho/) appears in many new sources, especially in electronic publications [Rŭzgor 2011-04-27; Muhammad

---

[70] Generally speaking, the suffix -чӣ /-či/ seems to have been quite productive in the first half of the 20th century, as it is proven by numerous forms (not necessarily associated with scientific of technological development but also other fields of vocabulary), like: колхозчӣ /kolxoz-či/ 'a person employed in a Soviet-style collective farm', тракторчӣ /traktor-či/ 'a tractor driver', ленинчӣ /lenin-či/ 'a political follower of Lenin', сталинчӣ /stalin-či/ 'a political follower of Stalin'.

[71] NB the RU joining element /-o-/ is accidentally simillar to the one used in FA in stable coordinative phrases (TJ: /-u-/). However, as far as the RU one appears – as in the example above – in subordinate compounds, the use of the other is restricted to coordinate ones, e.g. обухаво /ob-u-havo/.

2011-03-16]. Of course, the original source of тонк is the same EN word **tank**. However, while in the case of танк we were entitled to suspect RU playing the role of an intermediary, the form тонк clearly shows DA/FA influence. The form تانک is used both in FA [Asadullaev & Peysikov 1965: 917; Omid 1373: 369] and in DA [Sādiqyār 1379 HŠ: 141; Ostrovskiy 1987: 352][72]. It is worth noting that while танк violates the standard TJ : DA/FA vowel correspondence /o/ : /ā/, the word тонк complies with it.

The form тонк is clearly less popular in online resources thank танк, as the search of this word together with the noun чанг /ʒang/ 'war' produces only below 500 results [Google search: keywords=тонк, чанг; date=2014-05-02] – c.f. above.

However, an interesting question is why Persian has /tank/ instead of */tank/, which would have been closer to the pronunciation of this internationalism in the languages that might have served as the immediate source of this form (English, French or even Russian). The answer to this question is that in spite of using different writing systems, Persian loanwords from European languages are often influenced by the orthographical form in their language of origin, e.g. the word فابریک 'factory' was taken from the FR **fabrique** (Omid 1373 HŠ: 897). The Persian loanword, in general, mimics the original pronunciation quite faithfully, with the exception of the vowel of the first syllable which is /ā/ even if the FA /a/ would have been phonetically closer to the original FR vowel (/a/). There are numerous examples of this phenomenon, like دیرژابل /diržābl/ instead of the expected *دیریژبل /dirižabl/, سلوفان /selofān/ instead of *سلوفن /selofan/, etc. In the case of تانک another important factor might have been a tendency to avoid ambiguity as potentially homographic forms exist (e.g. تنک /tonok/ 'narrow').

The TJ form тонк seems to be a newer one, as only one derivative has been found (with limited attestation), i.e. the adj. тонкӣ /tonk-i/ [Farhodi 2003][73], created with the adjectival suffix /-i/.

As the word **tank** is an internationalism, related forms are to be found in languages like PŠ تانک [Aslanov 1966: 212] or تانک /ṭānk/ [Lebedev et al. 1989: 675], TR **tank** [Alkım et al. 1996: 1094] and UZ/AZ/KY танк [Koščanov et al. 1984: 553; Alizade et al. 1944: 304; Yudahin 1957: 849].

As we see, most languages use the internationalism 'tank', however, there are exceptions – cf. PL czołg and AR دبابة [Ba'albaki 1999: 949].

---

[72] By the way, the form تانکیست 'tankman' is used in both FA and DA, too [Asadullaev & Peysikov 1965: 917; Ostrovskiy 1987: 352].

[73] Some hesitation in use of forms танкӣ and тонкӣ is visible in the cited source.

2.51 Tape recorder

The tape recorder was a result of a long process of inventions, all inspired by the need to record sound. Even reducing the perpective to devices recording on some kind of tape (thus, excluding wax cylinders etc.), the picture is quite complicated. The story starts with the non-magnetic recorders using a wax strip that were first constructed in the 19th century. However, modern type tape recorders using magnetic signal were introduced only in the 1930s. The AEG company introduced its first commercially successful model, i.e. the Magnetophon K1 in 1935, after building a number of laboratory recorders [Daniel et al. 1999: 56]. *AEG and BASF decided to call their type of sound recorder Magnetophon* [Daniel et al. 1999: 54].

The form used in TJ for the magnetic tape recorder is **магнитофон** [Osimi & Arzumanov 1985: 458; Myakišev & Buhosev 2000: 22] /magnitofon/. The history of this form may be traced to the trade name of the early tape recorder by the German firm AEG (**Magnetophon**). This DE form is a pseudo-classical compound of GR μάγνης, -ητος and φωνὴ.

The vehicular language was RU, where form identical with the TJ one exists (note esp. the vowel in the second syllable).

The RU form, in its turn, is a borrowing from DE, changed analogically to the already present in the language form **магнит**, which goes back to the same source, however, shows the Byzantine pronunciation of the vowel in the second syllable [Ward 1981: 6].

The word **магнитофон** is attested in TJ online resources [Ašůrov 2014-03-12; Jumhuriyat acc. 2014-09-10].

The plural form of **магнитофон**, i.e. **магнитофонхо** /magnitofon-ho/, is attested in TJ as well [Šikebo 2010-08-12]. A derived adjective **магнитофонӣ** /magnitofon-i/ with the suffix **-ӣ** /-i/ is attested [Jumhuriyat acc. 2011-03-27; Ӡůrayev acc. 2011-03-27; Maӡlis-i Oli 2009-11-19].

Another TJ form found is **мошинаи сабт** [Satskaya & Jamšedov 2000: 191] / mošina-i sabt/. As far as the head of this izofat phrase is concerned, see p. 46. The form **сабт** is of Arabic origin.

The phrase **мошинаи сабт** has not been found in online TJ resources [Google search: keywords= "мошинаи сабт", date=2014-09-15].

In FA, the most popular form is the native descriptive phrase دستگاه ضبط سوت / dastgāh-e zabt-e sowt/ 'a sound-recording device' [Klevcova 1982: 307; Asadullaev & Peysikov 1965: 378] (or simply ضبط سوت /zabt-e sowt/ for short). However, the form ماگنتوفون /māgnetofon/ is attested in FA, too [Asadullaev & Peysikov 1965: 378]. Apart from that, we find a form صدانگار [Google search: keyword=صدانگار, date=2014-05-12].

In DA we come across a loanword from EN, تیپ رکاردر /tēp rikārdar/ [Ostrovskiy 1987: 172; see also: Fishstein & Ghaznawi, 1975, 216] and تیپریکاردر /tēprēkārdar/ [Lebedev et al. 1989: 329]. In PŠ we find /ṭēprikārḍər/ [Penzl 1961: 45] or تیپ ریکاردر

/ṭiprikārdár/ [Lebedev et al. 1989: 329], and it is worth noticing that – like in some other instances – we have an isogloss here, encompassing PŠ and DA. However, another form is also used in DA, namely /nawār-zabt/ [Awde et al. 2002: 108].

The form TJ **магнитофон** finds parallels in other languages of the former Soviet Central Asia: UZ **magnitofon** [Balci et al. 2004: 145], KY **магнитофон** [Yudahin 1957: 337].

2.52 Television

*It is quite difficult to present one specific date of the invention of television. The story starts with the discovery of photoconductivity towards the end of the 19th century. The long quest for a system capable of scanning a picture, sending it to another place (by wire or radio) and reproducing it followed. The word **television** was coined in 1900 (see below), however, by that time it named a challenging project rather than a really existing technology. First electromechanical systems like that of Rignoux and Fournier (around 1906) were to some extant successful, but their functionality was still limited. It was only in 1925 that John Logie Baird demonstrated a transmission of images in motion [Burns, 2000, 85-86] and this is what we choose to be the date of invention of television as we understand it nowadays. Long-distance tv transmitions followed soon (1928).*

The form **телевизион** تلویزیون [Saymiddinov et al. 2006: 601; Nazarzoda et al. 2008: 2,338; Myakišev & Buhovsev 2000: 22] /televizyon/ is the dominating TJ version of the internationalism, the original source of which is the FR **télévision**. The FR form is a neologism – a compound of GR **τῆλε** "far" and Latin **vision** "view". It seems to have been used for the first time in a paper by Perskiy **телевизион** in 1900 and once it was introduced it replaced previously used terms like **telephot** or **telectroscope** [Abramson 1987: 23; Perskiy 1900: passim].

It is particularly interesting that it is the internationalism in question that is used in TJ and not the RU hybrid form **телевидение**. Unfortunately, it is difficult to indicate the vehicular language through which **телевизион** came into TJ.

The form is well attested in TJ websites [Siddiqšoh 2014-06-09; Asia-Plus 2013--08-07; Tojnews 2013-07-12].

In TJ an adjective derived from **телевизион** with the suffix **-й** /-i/, i.e. **телевизионй** /televizyon-i/ exists, too [Nazarzoda et al. 2008: 2,338; Myakišev & Buhovsev 2000: 135].

In FA, the most popular form is تلویزیون [Rubinčik 1970: 1,395; Google search: keyword=تلویزیون, date=2011-05-15], even if slightly different forms like تلویزیان [Google search: keyword=تلویزیان, date=2011-05-15] or تلویزان [Google search: keyword=تلویزان, date=2011-05-15] sometimes appear. In both DA and PŠ we come across تلویزیون /tiliwīzyún/ [Lebedev et al. 1989: 677], /teliwizyun/ [Awde et al. 2002: 67].

It is worth noting that traditional TJ-FA vowel correspondence rules are not obeyed in the case of **телевизион** and تلویزیون.

A similar form is attested in AR (beside forms like التلفزة etc. [Baʻalbaki 1999: 955]), i.e. التلفزیون [Baʻalbaki 1999: 955].

The form **телевизон** [Růzgor 2011-03-01; Prezident-i Toǯikiston 2010-06-26] /televizon/ is a less frequent variant of **телевизион** (an Internet search of **телевизон** within .tj domains produces less than 800 results [Google search: keyword=телевизон, domain=.tj, date=2011-05-15], while looking for **телевизион** we find over 400,000 pages [Google search: keyword=телевизион, domain=.tj, date=2011-05-15]). The difference between the form in question and the one discussed previously is not very significant and it is restricted to the final part of the word. The most plausible explanation of appearance of **телевизон** is probably the FA influence, as in the latter idiom the parallel form تلویزان is attested as well, even if not very popular [Google search: keyword= تلویزان, dates=2011-05-15, 2014-05-29]. Most probably the emergence of **телевизон** and تلویزان is a result of the reinterpretation of the loanword. After eliminating the glide from the last syllable, the word looks like one of the numerous PrsS ending in /-z/ (cf. **овез-** /ovez-/, **рез-** /rez-/, **гурез-** /gurez-/ etc.) with the participial suffix **-он** /-on/ attached (**овез-он** /ovez-on/, **гурез-он** /gurez-on/, etc.). Further derivatives based on **телевизон** exist, like the adj. **телевизонӣ** /televizon-i/ [Hizb-i Nahzat 2011-05-13; Qayumzod 2008-10-10].

A somewhat poetical form **оинаи нилгун** lit. 'blue screen (or blue mirror)' is also used in TJ [Saymiddinov et al. 2006: 405; Satskaya & Jamshedov 2000: 214; Radyo-i Ozodi 2005-11-30].

Another form, **садову симо** [Dostiyev 2009; Hamidov acc. 2011-05-17] /sado-vu simo/ seems to be a borrowing from FA, where we find صدا و سیما. In TJ it is not an extremely popular form, still an Internet search reveals over 1000 pages containing this form [Google search, keyword="садову симо", date=2011-05-17]. This form is particularly interesting as it is an example of coordinative stable phrase, which is quite rare in the analysed corpus.

Apart from the forms for 'television' as the whole system, it is interesting to mention one of the words used in TJ for 'tv set', i.e. **ҷаҳоннамо** [Saymiddinov et al. 2006: 721; Satskaya et al. 2007: 112].

2.53 Traffic lights

Although various attempts to control the traffic have been made as early as in the 18th century (related to horse driven carriages by that time), modern traffic lights appeared only in 1914 in Cleveland [Gazis 2002: 101].

The most popular TJ equivalent of the EN expression **traffic lights** is **чароғаки роҳнамо** [Saymiddinov et al. 2006: 701; Lutfulloyev 2002: 108-109; Boinazarov et al. 2007: 6] /čaroġak-i rohnamo/. It is based on elements previously existent in the lexical system of TJ. Basically, it is a stable izofat phrase, the first (descripted) element of which is the noun **чароғак** /čaroḡak/ (n. **чароғ** /čaroḡ/ 'a lamp' + suffix -**ак** /-ak/), the other being the form **рохнамо (рахнамо)**. The latter itself is a compound, consisting of the noun **рох/рах** 'a way, road' and the present stem **намо** /namo-/ of the verb **намудан** /namudan/ 'to show'.

The phrase is attested in online TJ resources [VKD 2013-04-22; Mirbozxonova 2014-05-16], as well as its plural form, **чароғакхои рахнамо** /čaroḡak-ho-i rahnamo/ [Axtar 2011-03-28].

As far as its origin is concerned, it does not seem to be a simple calque of forms used in RU, EN or other languages of the West. However, its similarity to the FA term (see below) entitles us to ask, if it is not a borrowing from that closely related idiom involving additional adaptation.

The FA parallel form is چراغ راهنما /čerāq-e rāhnamā/ [Āryānpur & Āryānpur 1375: 394; Klevtsova 1987: 616; Asadullaev & Peysikov 1965: 820.]. It differs from the TJ form in that its first element does not have the suffix /-ak/.

In DA, we find a related form, too: چراغ ترافیکی /čerāḡ-e tarāfiki/ [Ostrovskiy 1987: 322; Wahab 2004: 175] or /čarāḡ-e trāfīkī/ [Lebedev et al. 1989: 629]. The difference between the FA and TJ forms on the one hand, and the DA one on the other is the second element of the izofat phrase. Instead of the compound /rāh-namā/, an adjective derived from the loanword /tarāfik/ (← EN **traffic**) by adding the suffix /-i/. Another form used in DA is /ešāre-ye tarāfiki/ [Awde et al. 2002: 110].

In PŠ we find a related phrase, too, namely ترافیک څراغ د /də trāfík crāḡ/ [Lebedev et al. 1989: 629].

Another form used to denote 'traffic lights' is **светофор** [Bertel's et al. 1954: 352; Baizoyev & Hayward 2003: 356; Bashiri 1994: 127] /svetofor/. It is a loanword from RU, and it is not analysable word-formationally on the ground of the TJ language. The same form is attested e.g. in KY: **светофор** [Yudahin 1957: 761] and in AZ: **svetofor** [Öztopçu 2000: 380].

The plural form **светофорхо** /svetofor-ho/ is used in TJ as well [Nabi acc. 2011-03-28; Axtar acc. 2011-03-28].

Another form used (even if not really popular) is the compound **сечароғ** [Moukhtor et al. 2003: 227; Mŭso 2012-04-05]. It consist of the numeral **се** 'thee' and the noun **чароғ**, so the literal translation would be 'three lamps', or better: 'something that has three lamps' (a *bahuvrīhi* compound). The form is known in FA (in the form سه چراغ), as well and it might have been borrowed from that idiom.

2.54 Transistor

The first working semiconductor transistor was demonstrated in 1947 [Ellinger 2008: 113]. *This term was coined in 1949 by John Pierce at BTL* [Yeo et al. 2010: 2] *as a compound of unmorphologically abbreviated **trans(fer)** + **(re)sistor*** [See Spirkin et al. 1980: 514].

The word **транзистор** [Saymiddinov et al. 2006: 610; Myakišev & Buhovsev 2000: 75; Rahimov et al. 2006: 221] /tranzistor/ is a typical internationalism. It comes from EN (v. sup.), whence it came into RU [Spirkin et al. 1980: 514]. Though it is difficult to indicate any decisive proof, we may put forward the hypothesis that the word was transferred from EN to TJ via RU (like **vitamin** and many other).

The form is attested, albeit not in great numbers, in online materials [Usmoniyon 2012-04-30; Shuhratjon 2014-01-22].

The word **транзистор** is used in izofat phrases like **радиои транзистор** [Salihov 1990: 69].

The plural form **транзисторхо** /tranzistor-ho/ is used as well [Myakišev & Buhovsev 2000: 18; DMT 2010-06-07] and so is the derived adj. **транзисторӣ** / tranzistor-i/ with the suffix **-ӣ** /-i/ [Rahimov et al. 2006: 221; Ӡonmard 2010-01-22]. A derivative with prefix: **батранзистор** /ba-tranzistor/ 'with a transistor, containing a transistor' is attested, too [Myakišev & Buhovsev 2000: 75].

The same internationalism is attested in FA & DA: ترانزیستور /t(e)rānzistor/ [Āryānpur & Āryānpur 1375: 289; Rubinčik 1970: 1,363; Haghshenas et al. 2002: 1810], /tarānzīstōr/ [Ostrovskiy 1987: 358] or /trānzīstōr/ [Lebedev et al. 1989: 689]. In PŠ the form ترانزیستور /trānzistór/ is attested [Lebedev et al. 1989: 689].

In UZ we find **транзистор** [Koščanov et al. 1984: 585] as well. The same internationalism is used, e.g. in AR: الترانزیستور [Ba'albaki 1999: 984].

2.55 Transparent Adhesive Tape

Transparent adhesive tape (known under the marketing name Scotch tape) was invented in 1930 [Cole et al. 2003: 22].

The internationalism based on the marketing name Scotch is used in TJ, like in many other languages: **скотч** [Čarx-i Gardun 2011-01-20; Khovar [2011]-07-21]. No derivatives of this form were found. As there seems to be more tolerance for initial consonant clusters in TJ than FA (owing to the RU influence), the structure of the word is not a decisive argument for classifying it as a foreign word rather than a loanword. Nevertheless, such a hypothesis may be supported by the fact that it is also morphologically poorly assimilated – no derivatives or plural forms have been found.

A parallel form, differing only in the structure of the first syllable, is used as the describing part in the izofat phrase used in FA, namely چسپ اسکاچ [Haghshenas et al. 2002: 1517] /časp-e eskāč/.

The same internationalism is also used in RU and in other languages of former Soviet Central Asia, e.g. UZ **ско(т)ч** [tib.islom.uz 2011-04-23; Abdurahmonova 2010-10-21]. Thus, it is justified to assume that RU played the role of a vehicular language here.

Another form used in TJ is **когази сиреш** /qoḡaz-i sireš/ [Khovar [2011]-07-21], where **сиреш** / **ширеш** stands for 'glue'. This, however, does not seem to be very popular and even in the cited source it had to be glossed by the form **скотч**.

Apart from the form mentioned above, at least two more phrases are used in FA referring to adhesive tape نوار چسپ and چسپنواری [Haghshenas et al. 2002: 1517].

2.56 Vacuum tube

The first vacuum tube was patented by Fleming in 1904 [Kragh 2002: 127].

The TJ equivalent of the EN term 'vacuum tube' is an izofat phrase **лампаи электронӣ**, with the loanword **лампа** /lampa/ as the head, and the adjective **электронӣ** /elektroni/ 'electronic', itself created by suffixation (adj. suffix /-i/) from the noun **электрон** /elektron/. The form is attested in a very limited number of websites [Google search: keywords= "лампаи электронӣ", date=2014-09-16].

The word **лампа** 'a lamp' is another internationalism borrowed into TJ from RU, as a rival to the native **чароғ** [Bashiri 1994: 118]. To RU, in its turn, it was taken from FR (just like into EN and DE), and it has been attested there since the 17th c. [Černyx 1999: 1,465]. The source of the FR **lampe** is the LA **lampas** and the latter was borrowed from GR **λαμπάς** 'a torch' [Ibid.]. For the origin of the form **электрон** see **почтаи электронӣ** (p. 46).

A similar form is attested in FA lexicographical sources: لامپ الکترونی [Argāni 1364:180] /lāmp-e elektroni/. An interesting difference between the TJ & FA forms is the final /-a/ of the word previous one, which indicates RU as the immediate source. The FA variant is, understandably, deprived of this element. Moreover, there is no regular phonetic correspondence between the two forms, as the TJ /a/ (and not /*o/) is the counterpart of the FA /ā/ in this case.

The TJ form is attested in the lexicography [Bertel's et al. 1954: 203; Nazarzoda et al. 2008: 1,711] and in other sources [Rahimov et al. 2006: 224; Mažidov & Nozimov 2006: 48; Farzin 2008-08-01].

The plural form **лампахои электронӣ** [Šarifov acc. 2012-12-30] exists.

In FA, other forms, apart from لامپ الکترونی, are attested, e.g. لامپ الکترونیک [Gyunašvili 1974: 186], چراغ الکترونیکی [Mirzabekyan 1973: 288], لامپ ترمویونیک [Mirzabekyan 1973: 288], لامپ رادیونی [Gyunašvili 1974: 186], لامپ خلأ [Haghshenas et al. 2002: 1881; Argāni 1364: 180] /lāmp-e xalā'/, لوله خلأ [Haghshenas et al. 2002: 1881] or – a somewhat more elaborated form – لامپ خلأ شده [Argāni 1364:534].

As far as the other languages of the region are concerned, we find the UZ **электрон лампа** [Koščanov et al. 1984: 762], KY **электролампа** and **электр лампасы** [Yudahin 1957: 973].

2.57 Vitamin

*The noun **vitamin** was introduced by Kazimierz Funk in 1912* [Rembieliński & Kuźnicka 1987: 132; Messadié, 1995, 234].

Once introduced in the second decade of the 20th century, the form **vitamin** became one of the wide-spread internationalisms. It was coined in EN by Funk (v. sup.), and it is based on the Lat. **vita** 'life' and **amine** (hence the original spelling **vitamine**) 'type of organic chemical compounds' ← Lat. **ammoniacum** [see Tokarski et al. 1980: 28, 808]. Черных traces presence of this form in RU to the first translation of Funk's book Vitamins which took place in 1922 [Černyx 1999: 1,154]. The TJ form **витамин** ويتمين /vitamin/ is attested in numerous sources [Bertel's et al. 1954: 92; Moukhtor et al. 2003: 40; Saymiddinov et al. 2006: 130; Nazarzoda et al. 2008: 1,277; Bashiri 1994: 122] (together with its plural form **витаминхо** /vitamin-ho/ [Osimi & Arzumanov 1985: 106]). The most probable vehicular language in the case of this internationalism seems to be RU.

Compounds and affixal derivatives with the form in question as their element/ base are attested, e.g.: **витаминдор** ويتمين‌دار /vitamindor/ 'containing (a) vitamin(s)' [Nazarzoda et al. 2008: 1,277; Osimi & Arzumanov 1985: 106], where the second element (the head) is the PrsS **дор-** /dor-/ of the verb **доштан** /doštan/ 'to have, to possess'. There is also a further derivative based on the latter, i.e. **витаминдорӣ** [Bertel's et al. 1954: 93] 'possessing vitamins'. Another example is the adj. **витаминӣ** ويتمینی /vitamin-i/, the meaning of which is in fact synonymic to **витаминдор** [Nazarzoda et al. 2008: 1,277; Osimi & Arzumanov 1985: 106] and which is built by adding the adj. suffix -**ӣ** /-i/ to the base form. An adjective derived from **витамин** / vitamin/ by the means of the prefix **бе-** /be-/, i.e. **бевитамин** /be-vitamin/ 'deprived of vitamins, having no vitamin' exists, too [Gulxoja 2008]. Among forms of this sort, it is worth mentioning the word **камвитамин** /kam-vitamin/, the status of the first element of which is disputable[74].

Related forms exist both in FA: ويتامين [Āryānpur & Āryānpur 1375: 1388; Asadullaev & Peysikov 1965: 92] and DA: ويتامين /witāmin/ [Sādiqyār 1379 HŠ: 177; Ostrovskiy 1987: 57; Fishstein & Ghaznawi 1979: 243]. What is worth noticing is the fact that the sound correspondences between TJ, FA & DA are not obeyed here that is we have **витамин** instead of the expected *****витомин** (cf TJ **роёна**: FA رايانه – see the section 'Computer').

Other languages of the region use the same loanword: PŠ ويتامين /witāmín/ [Lebedev 1961: 89; Lebedev et al. 1989: 103], UZ **vitamin** [Balci et al. 2004: 303],

---

[74] A number of forms like **кам** /kam/, **пур** /pur/, **сохиб** /sohib/ occupies an intermediary position between prefixes and elements of compounds, i.e. affixoids [See Bussman 1998: 25]. This phenomenon appears in all three varieties of Persian. Рубинчик in his grammar of FA calls them "semi-affixes" [Rubinčik 2001: 148ff]. If it is a justifiable practice, surely the forms like the mentioned ones are the best candidates to be classified like that [See Rubinčik 2001: 153]. Looking at this problem from the diachronic point of view, we may see these forms as affixes in statum nascendi [Bussman 1998: 25].

KY **витамин** [Yudahin 1957: 81]. The internationalism in question is very expansive and its form may be found probably in most of the languages of the world. Equivalents based on local lexical corpus are rare, note, however, AR الحيمين [Arslanyan & Šubov 1977: 106] beside الفيتامينون Arslanyan & Šubov 1977: 106].

2.58 Webcam

The first webcam was constructed in the Cambridge University computer laboratory in 1991 [Gross et al. 2003: xxii].

The TJ word for a 'webcam' is **вебкамера** /veb-kamera/. To the author's best knowledge, it has not been yet included in lexicographical works, nevertheless, it has been attested in other (mostly on-line) sources since at least 2005 [Gacek 2007: 19]. It is an internationalism, however, as its form is identical with one of the most popular ones attested in RU, we are entitled to suppose that the latter language served as an intermediary here. This even more probable, if we consider the fact that the original form (**web camera**) is today much less popular in EN than the abbreviated from **webcam**, while in RU, the borrowing based on the previous one is used.

The plural form **вебкамерахо** is attested [Ruhulloh 2012-07-01], even if it is not very popular.

The origin of the first element of the EN compound **webcamera** may be traced to the GR **καμάρα** 'an arched room', which passed via LA **camera** [cf. Tokarski et al. 1980: 333] to various European languages, EN included.

The plural form **вебкамерахо** is attested, too [Ruhulloh 2012-07-01].

In FA we find a variety of forms including ones resembling the TJ **вебкамера**, like وبكم /vebkam/ & وبكام /vebkām/ [Gacek 2007: 19]. Apart from that, however, we find forms based (at least) partly on indigenous lexemes, like رایابین /rāyābin/ [Gacek 2007: 19] and وببین /veb-bin/ [Wikipedia: entry=وببین , date=2011-01-25]. Apart from that, the EN form **webcam** in the Latin script is used in FA texts, too [Gacek 2007: 19].

2.59 World Wide Web

The first working www server and browser started working in 1990 [Macnamara 2010: 47].

The loanword **веб** is used in TJ [Atoyev acc. 2011-10-06: 1]. It also appears in a large number of derived words like **веб-сомона** [VKD acc. 2011-10-06], **веб-блог** [Baxtiyor 2009-10-06], **веб-сайт** [Nekrůzov [2010]-10-13] or **вебкамера** (q.v.).

The same internationalism is present e.g. in AR: الويب [Google search: keyword=الويب , date=2011-10-06], where it is used as the basis for various collocations and further derivatives like مقهى الويب العربي 'Arab web-caffee' [Arabwebcafe 2011-08-28] etc.

Another TJ form for the World Wide Web attested is **туранкабути чаҳонӣ** [Atoyev acc. 2011-10-06: 1]. This form is very rare (possibly a hapax) but still it is very interesting, as it contains a word for"spider", which is not present explicitly in the EN equivalent. Cf. AR الشبكة العنكبوتية العالمية [Elshami [2010-11-12]; Deutsche Welle 2008-04-30].

In FA we find forms like: جهان وب [Abbāsi 1387-06-27 HŠ; Google search: key-word= جهان وب, date= 2014-08-15] or simply وب [Haghshenas et al. 2002: 1923]. Another phrase used in FA to refer to WWW is وب جهان گستر [Aftabir 1386-04-16 HŠ; Google search: keywords= "وب جهان گستر", date= 2014-09-27]. Similar forms like وب‌جهان‌شامول and تارجهان‌گستر are attested as well [Zare'i 1389-09-10]. Apart from that, one may also mention شبکه جهانی [Haghshenas et al. 2002: 1923].

DOI: 10.12797/9788376385310.03

3 Conclusions

Even if the quantity of forms analysed in the present work (over 150 entries) is probably not large enough to put forward decisive conclusions on the modern TJ vocabulary as a whole, nevertheless it allows to formulate a number of valuable observations.

First of all, let us focus on the methods of enriching the vocabulary noticed in the studied forms. As the vocabulary analysed in the present work refers to objects and phenomena non-existent or unknown before the beginning of the 20th century, they are all new, at least in the sense of connection between a given lexeme and an object or phenomenon. Singular exceptions are forms like helicopter or robot (the later not included in this work), where some concept was imagined or predicted before the actual invention or discovery. Their number is, however, not significant enough to change the overall image.

Let us pay particular attention to the processes leading to appearance of new forms on the basis of the pre-existent lexemes. There is no doubt that with such an assumption the most productive mechanism is *izofat*. As a consequence, forms used to name new inventions/discoveries are mostly phrases rather than words. Borrowing is the second most popular mechanism used to enrich the TJ vocabulary. However, even counting borrowings together with foreign words, their number is still lower than that of izofat phrases. The number of cases where word-formational mechanisms (composition, affixation, sentence-petrification and clipping) are employed is even lower.

The number of processes applied to create the analysed forms on the basis of already existing ones (or – in the case of borrowings and foreign words introduction – on the basis of a foreign form) varies from one to three. The vast majority (more than 130 forms) needed only one process, while around 20 were created in two steps. Less than ten forms required three steps to be constructed on the basis of pre-existent elements.

In the case where two or three processes take place, it is mostly an instance of two (10 examples) or three (below 5 examples) consecutive izofat constructions, e.g. **диски миқнотисии чандирӣ, бемории норасоии муҳассали масуният**. Izofat may also be combined with other mechanisms, as borrowing (**телефони мобайл**) or suffixation (**бомбаи кассетавӣ**). One example has been found, where izofat co-exists with borrowing and suffixation (**телефони мобилӣ**). An instance where the modifier of an izofat phrase is a foreign word has been found, too (**хисобгараки Гейгер-Мюллер**).

Most of the izofat constructions in the analysed corpus (more than 50 instances) consist of a noun being the head of the phrase and and adjective playing the role of the modifier, e.g. **радифи маснӯъ**, **бомбаи атомӣ**, **суратгираки дичитол** etc. Nominal izofat phrases (i.e. the ones with a noun serving as the modifier, e.g. **аломати норасой**, **асбоби шунавой** and so on) are less common (less than 30 examples).

The form **SMS-паёмак** may be classified as created by foreign word introduction and suffixation followed by composition. Apart from that, suffixation may accompany composition (**чангкашак**) or sentence petrification (**мӯхушккунак**). Another specific example is the word **хавопаймо,** which was first created as a compound to be later influenced semantically by the parallel FA form.

In the case of the borrowings and foreign words, it is interesting to study their sources, especially the original source and the immediate source (i.e. the last of the vehicular languages). Although, establishing these facts may be disputable in some cases, still we are able to make some observations. Namely, EN is the most common original source and over 50% borrowings/foreign words that were possible to classify on the basis of this factor are ultimately derived from this language. Interestingly, the position of RU is not particularly significant here, as statistically it is on the same level as FA, DE and FR. Only isolated examples of foreign words take from RU has been found, e.g. the acronym **СПИД** and the noun **огнемёт**.

FA seems to be the original source of about ten of the analysed forms. Even if the borrowings from FA are not the most prominent group, still they deserve some attention. In most cases these forms appeared quite recently, even though a synonymical form had existed before, e.g. **тонк** (c.f. **танк**), **хурмун** (c.f. **гормон**), **резпардозанда** (c.f. **микропротсессор**) etc. This phenomenon is particularly important, as it is a proof that some tendencies towards maintaining (renewing) the ties between TJ and the FA of Iran are still/again present. It may also be proven that these forms gain popularity as the time goes on, e.g. the form **роёна** (q.v.) was almost a hapax in 2004 and had to be glossed when used in a TJ text [Gacek 2007: 24]. Today, even if it is not the first-choice word for 'computer', it appears in hundreds of websites.

Analysing these results, we have to keep in mind that the original source is understood in the present work, as the language, in which a word was used for the first time with the meaning discussed here. However, the form in question might have been created by word-formational mechanisms, coined of foreign elements, borrowed with semantic change involved etc. Thus, the form (or its elements) itself may have much longer history. A good example is provided by the word **аэроплан** (q.v.). Its original sources is FR, as it is there that the form **aéroplan** was coined. However, as one of the numerous pseudo-classical compounds in FR, it was built of GR elements ἀήρ and πλάνος. Thus, even though neither GR or LA appear as an original source of a single form analysed in the present work, still we find among these forms quite a lot of words containing elements, the history of which may be traced back to these languages (e.g. **телевизион, аллергӣ, антибиотик**, etc.)

Let us now pay attention to the problem of the immediate sources. Here RU is unquestionably the winner with more than half of the forms for which the vehicular language was indicated. It should be noted is that RU plays this role in the case of some forms originated in EN. Many of these forms are acronyms (e.g. **AIDS, SMS**) or partially acronyms (**MP3-плеер**). Other are derived from proper names, e.g. **скоч, Bluetooth**. in the case of some most recent borrowings with original sources in EN the possibility of direct borrowing may not be excluded, even if RU still seems to be highly possible as a vehicular language.

We notice that many of the ideas discussed in this book may be named in TJ using a number of different forms, belonging to different layers of vocabulary, defined on the basis of the lexical forms origin. Probably the best example of this specific form of synonymy is the idea of 'an airplane'. There are four forms used to convey the idea of 'an airplane' in TJ. The first is the FR **аэроплан**, most probably transmitted into TJ via RU. Also the native RU **микропротсессор** form **самолёт** is used. The form **тайёра** is derived from AR, while **ҳавопаймо** came to designate an airplane under the influence of FA. In most cases the number of synonyms belonging to different layers of vocabulary is much more limited, c.f. the adj. **хаста(в)й** and the parallel form of RU origin, i.e. **ядрой** or **резпардозанда** and **микропротсессор**.

Apart from borrowings and foreign words, we notice other types of the influence of other languages on the lexical system of TJ. We find a number of language calques in the analysed material, however, in many cases it is difficult to name the particular source of inspiration, as we observe a tendency that certain forms tend to be calqued rather than borrowed in many languages, e.g. black box (cf. DE **schwarzer Kasten,** RU **чёрный ящик**, PL **czarna skrzynka)**.

Different types of foreign impact on TJ vocabulary include semantic borrowings especially from FA, the most striking example being **ҳавопаймо**. Apart from the change of meaning, some forms changed their phonetic form or developed phonetic variants definitely or probably under the influence of this sister-language of TJ (e.g. **тонк** besides **танк, бомб : бомба, машина** > **мошина**). Taking all that into consideration together with the borrowings from FA present in TJ (see above), we have to say that the influence of Persian – as it is spoken in Iran – on the Tajik is still observable even in the most modern layers of the vocabulary.

Another sphere that deserves some attention is a number of orthographic phenomena. Despite the fact that a tendency to return to the Perso-Arabic script was strong in Tajikistan in 1990s and it was legally accept **микропротсессор** ed, still Cyrillic alphabet remains the only writing system universally used to write TJ. However, some changes (a low-scale orthography reform) were introduced in that consisted mostly of letters thought to be of no use in TJ, as opposed to RU, i.e. **микропротсессор** ц, щ, ы and ь. It may be interpreted as a manifesto of the fact that the Stalin's rule demanding all the borrowings from RU in other languages of the Soviet Union to be written in accordance with RU orthography [see Perry 1997: 12].

Most of the analysed forms comply with this reformed Cyrillic orthography rules. The most recent words substitute the abandoned letters – whenever necessary – with equivalent letter sequences, e.g. **тс** is used instead of **ц** (e.g. **микропротсессор**). Interestingly, in some forms reveal specific phenomena of RU phonology that are not marked in their original form, see e.g. the word like **манитор**, **микропратсессор** (q.v.), showing the RU phonetic phenomenon of *akanye*, which is not marked in writing in its original orthography.

There is a tendency to use foreign (mainly EN) abbreviations in the Latin script (e.g. **SMS, CD**), sometimes creating hybrid, Latin-Cyrillic forms (e.g. **USB-хофиза**).

Finally, the question comes up if we are able to indicate any tendencies in the development of the analysed vocabulary? A number of phenomena observable in various analysed forms, namely re-borrowings from FA (e.g. **роёна**), changing the meaning of a form under the FA influence (**хавопаймо**), alternation of the phonetic form under the FA influence (**мошина**) have all one common feature – they may all be described as a tendency to re-iranization of vocabulary or – in other words – to retain (or regain) unity of TJ and FA, at least on some level. This certainly does not mean that RU is not appearing as an immediate source in the case of the most recent items of the analysed vocabulary. Nevertheless, the re-iranization tendency seems to gain momentum.

An interesting phenomenon which may be noted is an observable level of discrepancy between lexicographical works and heterogeneous online resources. It probably results from some kind of puristic tendencies on the side of the authors of modern TJ dictionaries. One may note, e.g. that the RU loanword **самолёт** (q.v.) does not appear in some modern works, while it may be found in other dictionaries together and a considerable number of TJ-language texts. Similarly, the form **холодилник** (q.v.) is totally absent from dictionaries, but it is – nevertheless – used in online electronic texts.

To sum up, research of the modern TJ lexicon reveals interesting factors within TJ lexical system that may be helpful in understanding directions of its development. As it has been stressed, certainly the scope of this book is vast enough to find decisive answers. Nevertheless, what we observe is a tendency to preserve (regain) unity of the Persian (sensu largo)-speaking world rather than further differentiation from Tajik from Fārsi. Certainly the problem deserves more effort on the side of Iranian studies specialists.

4 Alphabetical Index

AIDS	**2.1**
ATM	**2.7**
Bluetooth	**2.11**, 3
CD	**2.12**, 3
CD-ROM	**2.12**
E-mail	**2.27**
email	**2.27**
MP3-бозигар	**2.43**
MP3-плеер	**2.43**, 3
SMS	**2.48**, 3
SMS-ӣ	**2.48**
SMS-паём	**2.48**
SMS-паёмак	**2.48**, 3
аксбардораки рақамӣ	**2.20**
аллергӣ	**2.3**, 2.13, 3
аллергия	**2.3**
аллергиявӣ	**2.3**
аллергияовар	**2.3**
аломати мухассали масунияти одам	**2.1**
аломати норасоии масунияти бадан	1.8.7, **2.1**
АНМБ	**2.1**
антибиотик	**2.4**, 3
антибиотикӣ	**2.4**
асбоби шунавой	**2.32**, 3
аэроплан	**2.2**, 3
аэроплансозӣ	**2.2**
аэропланчӣ	**2.2**
банкомат	**2.7**
банкоматӣ	**2.7**
барфпок	**2.8**
барфпоккун	**2.8**, 2.13
бемории норасоии бадан	**2.1**
бемории норасоии масунияти одам	**2.1**
бемории норасоии мухассали масуният	1.8.7, **2.1**, 3
бемории пайдошудаи норасоии масуният	**2.1**
Блутус	**2.11**
бомбаи атомӣ	**2.6**, 2.15, 2.26, 2.45, 3
бомбаи кассетавӣ	**2.15**, 3
бомбаи кассетӣ	**2.15**
бомбаи кластерӣ	**2.15**
бомбаи хӯшай	**2.15**
бомбаи хаставӣ	**2.6**
бомбаи хастай	**2.6**, 2.45
бомбаи ядрой	**2.6**
бомби атомӣ	**2.6**
веб	**2.59**
веб-блог	**2.59**
веб-сайт	**2.59**
веб-сомона	**2.59**
вебкамера	**2.58**, 2.59
вертолёт	**2.33**
витамин	**2.57**
бевитамин	**2.57**
витаминдор	**2.57**
витаминдорӣ	**2.57**
витаминӣ	**2.57**
камвитамин	**2.57**
гардкашак	**2.24**
гитараи электрикӣ	**2.21**
гормон	1.3, **2.34**, 3
гормондор	**2.34**
гормонӣ	**2.34**
гурӯхи хун	**2.10**
гурӯхи хунӣ	**2.10**
Дандони обӣ	**2.11**
дастгохи дурӯғсанҷ	**2.39**
дастгохи майкровейв	**2.41**
дастгохи рақамии суратгирӣ	**2.20**
дастгохи ташхиси дурӯғ	**2.39**

дастгоҳи худпардоз	2.7	нейтронборонкунӣ	2.44
детектори дурӯғ	2.39	нейтронӣ	2.44
дискет	2.29	нерӯгоҳи атомӣ	2.45
дискета	2.29	нерӯгоҳи баркии атомӣ	2.45
диски магнитии чандирӣ	2.29	нерӯгоҳи ҳаставӣ	2.45
диски миқнотисии чандирӣ	2.29, 3	нерӯгоҳи ҳастай	2.45
дурӯғсанҷ	2.39	норасоии масуният	2.1
ЕЙДЗ	2.1	огнемёт	**2.28**, 3
заррабини электронӣ	2.26	оинаи нилгун	2.52
изотоп	2.37	оташандоз	2.28
изотопӣ	2.37	оташдони микромавҷ	2.41
инсулин	2.35	оташпош	2.28
инсулинӣ	2.35	оташпошанда	2.28
калкулятор	2.25	паёмак	2.48
компакт-диск	2.12	паёми кутоҳ	2.48
компутер	2.16	печка[и] микромавҷ	2.41
компютер	2.16	пости электронӣ	2.27
компютерӣ	2.16	почтаи электронӣ	**2.27**, 2.56
компютеркунӣ	2.16	почтаи электроникӣ	2.27
қоғази сиреш	2.55	прион	2.47
лазер	2.38	радифи маснӯӣ (Замин)	2.5
лазерӣ	2.38	радифи маснӯи Замин	2.5
лампаи электронӣ	2.56	радифи маснӯъ	2.5
магнитофон	2.51	радифи маснӯъи Замин	2.5
магнитофонӣ	2.51	радифи маснӯъии Замин	2.5
манитор	**2.42**, 3	резпардозанда	**2.40**, 3
микропратсессор	**2.40**, 3	ришгирак	2.23
микропротсессор	**2.40**, 3	ришгираки баркӣ	2.23
микропротсессорӣ	2.40	ришгираки электрикӣ	2.23
микроскопи электронӣ	2.26	роёна	2.8, 2.9, **2.16**, 2.19, 2.57, 3
микросхемаи интегралӣ	2.36	садову симо	2.52
монитор	2.42	самолёт	1.5, **2.2**, 3
моҳвора	2.5	самолётрон	2.2
моҳворабар	2.5	самолётронӣ	2.2
моҳворавӣ	2.5	самолётсозӣ	2.2
моҳворай	2.5	светофор	2.53
мошинаи гарду чанг ҷамъ кунанда	2.24	селлофан	2.14
мошинаи сабт	2.51	селлофанӣ	2.14
МП3-плеер	2.43	сечароғ	2.53
мӯйхушкунак	2.31	скотч	2.55
мӯхушккунак	2.31, 3	смс	2.48
муш	2.19	смс-й	2.48
мушак	2.19	СПИД	**2.1**, 3
мушвора	2.19	спутник	2.5
нейлон	2.46	спутникӣ	2.5
нейлонӣ	2.46	стансияи электрикии атомӣ	2.21, **2.45**
нейтрон	2.44	стеклоочиститель	2.8

суратгираки диҷитол	**2.20**, 3	ҳавопаймосозӣ	**2.2**
суратгираки рақамӣ	**2.20**	ҳамсафари сунъии Замин	**2.15**
схемаи интегралӣ	**2.36**	ҳассосӣ	**2.3**
тайёра	**2.2**, 3	ҳассосият	**2.3**
тайёрасозӣ	**2.2**	ҳеликуптар	**2.33**
танк	**2.50**, 3	ҳисобгараки Гейгер-Мюллер	**2.30**, 3
танкзан	**2.50**	ҳисобкор	**2.25**
танкӣ	**2.50**	ҳофиза	**2.18**
танкрон	**2.50**	USB-ҳофиза	**2.18**
танкронӣ	**2.50**	ҳуликуптар	**2.33**
танксоз	**2.50**	ҳурмун	**2.34**, 3
танксозӣ	**2.50**	ҳурмунӣ	**2.34**
танкчӣ	**2.50**	чангкашак	**2.24**, 3
танкшикан	**2.50**	чароғаки роҳнамо	**2.53**
телевизион	2.5, **2.52**, 3	чархбол	**2.33**
телевизионӣ	**2.52**	чархболбар	**2.33**
телевизон	**2.52**	чархболсозӣ	**2.33**
телевизонӣ	**2.52**	чип (*1. = схемаи интегралӣ*)	**2.36**
телефони дастӣ	**2.13**	чипдор	**2.36**
телефони мобайл	**2.13**, 3	чип (*2. = микропротсессор*)	**2.40**
телефони мобил	**2.13**, 3	ҷаъбаи сиёҳ	1.8.7, **2.9**
телефони мобилӣ	**2.13**, 3	ЭЙДЗ	**2.1**
телефони ҳамроҳ	**2.13**	яхдон	**2.22**
тонк	**2.50**, 3	яхдондор	**2.22**
тонкӣ	**2.50**	яхдонӣ	**2.22**
транзистор	2.54	яхдонсоз	**2.22**
батранзистор	**2.54**	яхдонсозӣ	**2.22**
транзисторӣ	**2.54**	яххона	**2.22**
тундпаз	**2.41**	яхчол	**2.22**
туранкабути ҷаҳонӣ	**2.59**		
фавкулноқил	**2.49**	آیرپلان	**2.2**
фавқуннокил	**2.49**	ائراپلن	**2.2**
файл	**2.17**	تلویزیون	**2.52**
файлӣ	**2.17**	چرخبال	**2.33**
фармавҷпаз	**2.41**	حسابکار	**2.25**
фен	**2.31**	ریزپردازنده	**2.40**
холодилник	**2.22**, 3	فوق الناقل	**2.49**
хотира	**2.18**	گارمان	**2.34**
худпардоз	**2.7**	لزیر	**2.38**
ҳавопаймо	**2.2**, 2.15, 2.5, 2.8, 3	نیترون	**2.44**
ҳавопаймобар	**2.2**	نیلان	**2.46**
ҳавопаймозадагӣ	**2.2**	هواپیما	**2.2**
ҳавопайморабо	**2.2**	ویتمین	**2.57**
ҳавопайморабоӣ	**2.2**	ویرتالیات	**2.33**

5 Bibliography

5.1 Form of bibliographical citations

The list of bibliography is divided into three subsections:
1. Linguistic sources (including works on lexicography). Generally, works included in this category are typical scientific publications. The only exception from this rule is a number of popular publications, which are, however, prepared in accordance with high standards, especially when it is guaranteed by their author(s) of publisher, e.g. a manual of Tajik by Hayword and Baizoyev (Hayword & Baizoyev 2003) or the RU-TJ phrasebook by Salihov (Salihov 1990).
2. Sources on the history of technology and science. These texts were used to establish the time and circumstances of particular discoveries and inventions. scientific sources were prefered if available, but common publications were also used.
3. Linguistic materials – texts in TJ (sometimes in other languages) documenting use of the analysed forms. These are not expected to be scientific works. Nevertheless, datable texts with clearly indicated authorship (or – at least – identity of the editor or publisher) were preferred.

Most of the sources used in the present work are published using one of the three writing systems: Arabic, Cyrillic and Latin, each of them in a number of variants. All the authors' names and titles were romanized to produce a coherent list, however, different methods of romanization were applied, depending on the original language and script. And so:
☞ Where a source proposes its own way of romanization of the publication data, this may be used, instead of the following methods;
☞ Bibliographical data of the sources published in Latin alphabet have been included in their original form, no matter what the language of the publication is. All language-specific characters are retained;
☞ Bibliographical data in the Arabic script are transcribed, rather than transliterated;
☞ For bibliographical entries in the TJ language written in the Cyrillic script, transcription is used;
☞ For all other languages using Cyrillic orthography (including Russian) transliteration is used;

Transliteration system used for RU:

Cyrillic letter	Transliteration sign
А	a
Б	b
В	v
Г	g
Д	d
Е	e
Ё	ë
Ж	ž
З	z
И	i
Й	y
К	k
Л	l
М	m
Н	n
О	o
П	p
Р	r
С	s
Т	t
У	u
Ф	f
Х	x
Ц	c
Ч	č
Ш	š
Щ	šč
Ъ	"
Ы	ỳ
Ь	'
Э	è
Ю	yu
Я	ya

This system looks complicated and incoherent at first, however, according to the author, it is probably the most ergonomical and practical one. Using one method universally would have go against traditional practices and create various problems. E.g. transliteration would have be impractical for Arabic-script publications, as only partial notation of vowels would have made bibliographical etries ambiguous and difficult to remember.

Similar methods are used in many publications (even if not described formally), e.g. by Perry (2005) and Rzehak (2001) q.v.

5.2 Linguistic works

Afanas'ev D. 1861. *Materialỳ dlya geografii i statistiki Rosii sobrannỳe oficerami genera-l'nogo štaba. Kovenskaya Guberniya.* Sankt Peterburg. Obščestvennaya pol'za.

Alizade M., et al. (eds.) 1944. *Kratkiy russko-persidsko-azerbaydžanskiy slovar'.* Baku. Izdatel'stvo Azerbaydžanskogo Filiala Akademii Nauk SSSR.

Alkım V. B., et al. (eds.) 1996. *Redhouse Yeni Türkçe-İngilizce Sözlük (16th edition).* İstanbul. Redhouse Yayınevi.

Argāni A. 1364 HŠ. *Farhang-e estelāhāt-e san'ati va fanni.* Tehrān. Entešārāt-e Amir Kabir.

Arslanyan G., Šubov Ja. 1977. *Russko-Arabskiy Mediceyskiy Slovar'.* Moskva. Izdatel'stvo Russkiy Yazỳk.

Āryānpur-Kāšāni A., Āryānpur-Kāšāni M. 1375 HŠ. *Farhang-e fešorde-ye fārsi be englisi.* Tehrān. Entešārāt-e Amir Kabir.

Asadullaev A., Peysikov L.S. (eds.) 1965. Russko-Persidskiy Slovar'. Moskva. Izdatel'stvo Sovetskaya Ènciklopediya.

Aslanov M.G. 1966. *Afgansko-Russkiy Slovar'.* Moskva. Izdatel'stvo Sovetskaya Ènciklopediya.

Awde N., et al. 2002. *Dari dictionary & Phrasebook. Dari-English/English-Dari.* New York. Hippocrene Books.

Ba'albaki M. 1999. *Al-Mawrid; A Modern English-Arabic Dictionary.* Beirut. Dar el-Ilm lil-Malayēn.

Bacon E.E. 1980. *Central Asians under Russian rule: a study in culture change.* London. Cornell University Press.

Baker M. (ed.) 1998. *Routledge Encyclopedia of Translation Studies.* London. Routledge.

Balci B., et al. 2004. *Nouveau Dictionnaire Ouzbek-Français.* Toshkent. Institut Français d'Études sur l'Asie Centrale.

Baizoyev A., Hayward J. 2003. *A Beginner's Guide to Tajiki.* London – New York. Routledge Courzon.

Bashiri I. 1994. *Russian Loanwords in Persian and Tajiki Languages* [in:] M. Marashi (ed.), *Persian Studies in North America. Studies in Honor of Mohammad Ali Jazayery.* Bethesda, Md. Iranbooks. pp. 109-39.

Bertel's E.È. et al. (eds.) 1954. *Tadžiksko-Russkiy slovar'.* Moskva. Gosudarstvennoe Izdatel'stvo Inostrannyx i Nacional'nyx Slovarey.

Burgmeier A., et al. 1991. *Lexis. Academic Vocabulary Study.* Englewood Cliffs, NJ. Prentice Hall Regents.

Bussman H. 1998. *Routledge Dictionary of Language and Linguistics* (translated form German by G. Trauth & K. Kazzazi). London – New York. Routledge.

Černyx P.Ja. 1999. *Istoriko-ètimologičeskiy slovar' sovremennogo russkogo yazỳka*. Moskva. Izdatel'stvo Russkiy Yazỳk.

Comrie B. 1981. *The Languages of the Soviet Union*. Cambridge – London – New York – etc. Cambridge University Press.

Danecki J., Kozłowska J. 1996. *Słownik arabsko-polski*. Warszawa. Wiedza Powszechna.

Dohlus, K. 2010. *The Role of Phonology and Phonetics in Loanword Adaptation: German and French Front Rounded Vowels in Japanese*. Frankfurt am Main. Peter Lang.

Dorofeeva L.N. 1960. *Yazỳk farsi-kabuli*. Moskva. Izdatel'stvo Vostočnoy Literatury.

Eddy A.A. 2007. *English in the Russian Context: A Macrosociolinguistic Study* (doctoral dissertation). Ann Arbor, MI. Proquest.

Eršov N.N., et al. 1942. *Voennỳy russko-tadžikskiy slovar'*. Stalinobod. Gosizdat Tadžikistana.

Ewing G. 1827. *Greek and English Lexicon*. Glasgow. University Press.

Fasmer M. 1987. *Ètimologičeskiy slovar' russkogo Yazỳka*. Moskva. Progress.

Fierman W., Garibova J. 2010. *Central Asia and Azerbaijan* [in:] J.A. Fishman & O. García (eds.) *Handbook of Language and Ethnic Identity. Disciplinary and Regional Perspectives*. Vol. 1. Oxford – New York – etc. Oxford University Press. pp. 423-469.

Filial-i Tožiki-i Akademiya-i Fanho-i SSSR. 1941. *Qoidaho-i Asosi-i Orfografiya-i Zabon-i Tožiki*. Stalinobod. Našriyot-i Davlati-i Tožikiston. (cited as: FTAF 1941).

Fishstein P., Ghaznawi M.R. 1975. *English-Dari Dictionary*. Kabul. Peace Corps (US).

FTAF – *see* Filial-i Tožiki-i Akademija-i Fanho-i SSSR.

Gacek T. 2007. *Computer Terminology in the Tajik Language* [in:] *Studia Etymologica Cracoviensia (SEC)*, vol. 12. pp. 17-29.

Gacek T. 2012. *Some Remarks on the Pronunciation of Russian Loanwords in Tajik* [in:] *Studia Linguistica Universitatis Iagellonicae Cracoviensis (SLING)*, vol. 129 supplementum. pp. 353-361.

Gacek T. 2014. *De-Russianisation of Internationalisms in the Tajik Language* [in:] *Studia Linguistica Universitatis Iagellonicae Cracoviensis (SLING)*, vol. 131. pp. 149-160.

Giunašvili Dž.Š. 1974. *Kratkiy Russko-Persidskiy texničeskiy slovar'*. Tbilisi. Metsniereba.

Goldsmith J.A., et al. (eds.) 2011. *The Handbook of Phonological Theory*. Publisher: Chichester, West Sussex – Malden, MA. Wiley-Blackwell,

Grenoble L.A. 2003. *Language policy in the Soviet Union*. New York – Boston – Dordrecht – etc. Kluwer.

Groves J. 1834. *A Greek and English Dictionary* (…). Boston. Hilliard, Gray and Company.

Haghshenas A.M., et al. 2002. *Farhang Moaser; English-Persian Millenium Dictionary*. Tehran. Farhang Moaser Publishers.

Hale A., Payton P. 2000. *New Directions in Celtic Studies*. Exeter. University of Exeter Press.

Hamzaev M.Ja. (ed.) 1962. *Slovar' Turkmenskogo Yazỳka*. Ašxabad. Izdatel'stvo Akademii Nauk Turkmenskoy SSR.

Haspelmath M., Tadmor U. 2009. *The Loanword Typology project and the World Loanword Database* [in:] M. Haspelmath, U. Tadmor (eds.), *Loanwords in the World's Languages: A Comparative Handbook*. Berlin. De Gruyter Mouton. pp. 1-34.

Haspelmath M., Tadmor U. (eds.) 2009. The World Loanword Database [at:] http://wold.livingsources.org/. Munich. Max Planck Digital Library (cited as: WLD 2009: entry name, access date).

Hindley R. 1990. *The Death of the Irish Language*. Abingdon, Oxfordshire – New York, NY. Routledge.

Householder F.W. jr., Lotfi M. 1965 *Basic Course in Azerbaijani*. Bloomington, IN – The Hague. Indiana University – Mouton & Co.

Ido Sh. 2005. *Tajik*. München. Lincom Europa.

Ido Sh. 2007. *Bukharan Tajik*. München. Lincom Europa.

Johnson L. 2006. *Tajikistan in the New Central Asia: Geopolitics, Great Power Rivalry and Radical Islam*. London – New York. I.B. Tauris.

Kalontarov Ja.I. 2007. *Farhang-i nav-i rusi-tožiki*. Dušambe

Kalontarov Ja.I. 2008. *Farhang-i nav-i tožiki-rusi*. Dušambe.

Kay G. 1995. English Loanwords in Japanese [in:] World Englishes, Vol. 14, No. 1. pp. 67-76.

Kerimova A.A. 1995. Ob osnovnyx processax razvitiya sovremennogo tadžikskogo litera-turnogo Yazỳka [in:] *Voprosỳ Yazỳkoznaniya. May-Iyun' 1995*. Rossiyskaya Akademiya Nauk. pp. 118-121.

Klevcova S.D. 1982. *Russko-Persidskiy Slovar'*: Učebnỳy. Moskva. Izdatel'stvo Russkiy Yazỳk.

Koptjevskaja-Tamm M. 2008. Approaching Lexical Typology [in:] M. Vanhove (ed.), *From polysemy to semantic change: towards a typology of lexical semantic associations*. Amsterdam. John Benjamins Publishing Company. pp. 3-52.

Koščanov M.K., et al. (eds.) 1983. *Russko-Uzbekskiy slovar'*, tom I. Taškent. Akademiya Nauk Uzbekskoy SSR.

Koščanov M.K., et al. (eds.) 1984. *Russko-Uzbekskiy slovar'*, tom II. Taškent. Akademiya Nauk Uzbekskoy SSR.

Kovtun L.S., Petuškov V.P. (eds.) *Slovar' Sovremennogo Russkogo Literaturnogo Yazỳka*, tom 17. Moskva – Leningrad. Izdatel'stvo Nauka.

Latify A.H. 1972. *Dari Newspaper Reader, vol. 1*. Washington. Foreign Service (Dept. of State).

Lebedev K.A. 1961. *Karmannỳy Russko-Arabskiy slovar'*. Moskva. Gosudarstvennoe Izdatel'stvo Inostrannyx i Nacional'nyx Slovarey.

Lebedev K.A., et al. 1989. *Russkiy-Puštu-Dari slovar'*. Moskva. Izdatel'stvo Russkiy Yazỳk.

Lorenz M. 1979. *Lehrbuch des Pashto (Afghanisch)*. Lepizig. VEB Verlag Enzyklopädie.

Malmkjær K. (ed.) 2010. *The Linguistics Encyclopedia*, Third Edition. London – New York. Routledge.

Megerdoomian K. 2009. Low-Density Language Strategies for Persian and Armenian [in:] S. Nirenburg (ed.) *Language Engineering for Lesser-studied Languages*. Amsterdam. IOS Press. pp. 291-312.

Mirzabekyan Ž.M. 1973. *Russko-persidskiy politexničeskiy slovar'*. Moskva. Izdatel'stvo Sovetskaya Ènciklopediya.

Misra G., et al. (eds.) 1995. *Deprivation: Its Social Roots And Psychological Consequences*. New Delhi. Concept Publishing Co.

Moinzadeh A. 2006. The Ezafe Phrase in Persian: How Complements are added to N°s and A°s [in:] *Journal of Social Sciences and Humanities of Shiraz University*, vol. 23, No. 1, pp. 45-57.

Moukhtor Ch. et al. 2003. *Dictionnaire tadjik-français*. Paris. Langues et mondes-l'Asiathèque.

Mühleisen S. 2010. *Heterogeneity in word-formation patterns: a corpus-based analysis of suffixation with -ee and its productivity in English*. Amsterdam – Philadelphia, PA. John Benjamins Publishing Company.

Muzofiršoyev M.O. 2009. Soxtor-i grammatiki-i navvožaho-i ismi (az rŭ-i mavod-i matbuot -i davri-i solho-i 90-umi sadda-i XX) [in:] *Payom-i Donišgoh-i milli-i Tožikiston. Baxš-i filologiya* 4(52). Dušambe. Sino. pp. 42-46.

Nazarzoda S., et al. 2008. *Farhang-i tafsir-i zabon-i tožiki* (iborat az 2 jild). Dušambe. Akademiya-i ilmho-i Ӡumhuri-i Tožikiston.

Nazarova S. 2011. Fa'oliyat-i Komissiya-i Baynihukumati-i Tožikiston-u Belarus gustariš meyobad [in:] *Payom-i Donišgoh-i milli-i Tožikiston* 10(74). Dušambe. Sino. pp. 255-257.

Olsen S. 2004. Coordination in morphology and syntax. The case of coordinate compounds [in:] A. Ter Meulen & W. Abraham (eds.) *The composition of meaning: from Lexeme to discourse.* Amsterdam – Philadelphia, PA. John Benjamins Publishing Company. pp. 17-37.

Omid H. 1373 HŠ. *Farhang-e Omid.* Tehrān. Entešārāt-e Amir Kabir.

Ostrovskiy B.Ja. 1987. *Karmannȳy Russko-Dari slovar'.* Moskva. Russkiy Yazȳk.

Osimi M., Arzumanov S.D. 1985. *Luḡat-i Rusi-Tožiki.* Moskva. Russkiy Yazȳk.

Öztopçu K. 2000. *Elementary Azerbaijani.* Santa Monica, CA – İstambul. L. Öztopçu.

Penzl H. 1961. Western Loanwords in Modern Pashto [in:] J*ournal of the American Oriental Society*, Vol. 81, No. 1 (Jan.-Mar., 1961). pp. 43-52.

Perry J.R. 2005. *A Tajik Persian Reference Grammar.* Leiden – Boston. Brill.

Perry J.R. 2009. *Tajik Persian* [in:] K. Brown, S. Ogilvie (eds.), Concise Encyclopedia of Languages of the World. Oxford. Elsevier. pp. 1041-1044.

Preobraženskiy A.G. 1958. *Ètimologičeskiy Slovar' Russkogo Yazȳka.* Moskva. Gosudarstvennoe Izdatel'stvo Inostrannyx i Nacional'nyx Slovarey.

Rastorgueva V.S., Kerimova A.A. 1964. *Sistema tadžikskogo glagola.* Moskva. Izdatel'stvo Nauka.

Rubinčik Ju.A. (ed.) 1970. *Persidsko-Russkiy slovar' v 2-h tomah.* Moskva. Izdatel'stvo Sovetskaya Ènciklopediya

Rubinčik Ju.A. 2001. *Grammatika sovremennogo persidskogo literaturnogo yazȳka.* Moskva. Vostočnaya Literatura.

Rudelson J.B. 1998. *Central Asia Phrasebook.* Hawthorn, Vic. Lonely Planet.

Rzehak L. 2001. *Vom Persischen zum Tadschikischen.* Wiesbaden. Reichert Verlag.

Sadeghi A.A. 2001. Language planning in Iran: a historical review [in:] *International Journal of the Sociology of Language.* Walter de Gruyter. pp. 19-30.

Sādiqyār M.A. 1379 HŠ. *Zabān-e Dari wa extelāt-e ān bā zabānhā-ye bēgāna.* Pēšāwar. Markaz-e Našarāti-ye Maywand.

Salihov S. 1990. *Muhovara-i Rusi-Tožiki.* Dušambe. Glavnaya Naučnaya Redakciya Tadžikskoy Sovetskoy Ènciklopedii.

Salihov S., Ismatullaev H. 1990. *Russko-Uzbeksko-Tadžikskiy razgovornik.* Taškent. Fan.

Saymiddinov D., et al. (eds.) 2006. *Farhang-i tožiki ba rusi.* Dušambe. Akademiya-i Ilmho-i Ӡumhuri-i Tožikiston.

Schlyter B. 2006. Changing Language Loyalties in Central Asia [in:] T.K. Bhatia & W.C. Ritchie (eds.), *The Handbook of Bilingualism.* Malden, MA – Oxford – Carlton, Victoria. Blackwell Publishing. pp. 808-834.

Skeat W.W. 1993. *Concise Dictionary of English Etymology.* Ware, Hertfortshire. Wordsworth Editions. (online: dx.doi.org/10.1093/acref/9780192830982.001.0001)

Sobirov È. 2007. Sud'bȳ rusizmov v tadžikskom yazȳke postsoveckogo perioda [in:] *Conference Abstracts. III Meždunarodnyȳ kongress issledovateley russkogo yazȳka. Russkiy Yazȳk: istoričeskiye sud'bȳ i sovremennost'.* 20-23 marta 2007 g. Moskva. pp. 147-150.

Spirkin A.G., et al. 1980. *Slovar' Inostrannyx Slov*. Moskva. Russkiy Yazẏk.

Stevenson A. (ed.) 2010. *Oxford Dictionary of English*. Oxford – New York – etc. Oxford University Press.

Summers D., et al. (eds.) 1995. *Longman Dictionary of Contemporary English*. Third edition. Essex. Longman Group Ltd.

Thomason S.G. 2001. *Language Contact*. Edinburgh. Edinburgh University Press.

Tokarski J., et al. (eds.) 1980. *Słownik wyrazów obcych*. Warszawa. Państwowe Wydawnictwo Naukowe.

Treffers-Daller J. 2010. Borrowing [in:] M. Fried, et al. (eds.), *Variation and Change: Pragmatic Perspectives*. Amsterdam – Philadelphia, PA. John Benjamins Publishing Company. pp. 17-35.

Wahab Sh. 2004. *Beginner's Dari (Persian)*. New York. Hippocrene Books, Inc.

Ward D. 1981. Loan words in Russian [in:] Journal of Russian Studies. Association of Teachers of Russian. 41 pp. 3-14; 42 pp. 5-14.

Yudahin K.K. (ed.) 1957. *Russko-kirgizskiy slovar'*. Moskva. Gosudarstvennoe Izdatel'stvo Inostrannyx i Nacional'nyx Slovarey.

5.3 Sources on the history of technology and science

Abramson A. 1987. *The history of television, 1880 to 1941*. Jefferson, NC. McFarland.

Achard K. 1989. *History and Development of the American Guitar*. Westport, CT. Bold Strummer.

Audoin C., Guinot B. 2001. *The measurement of time: time, frequency, and the atomic clock*. Cambridge – New York. Cambridge University Press. (online: dx.doi. org/10.1088/0957-0233/13/2/705)

Axelson J. 2005. *USB complete: everything you need to develop custom USB peripherals, 3rd edition*. Madison, WI. Lakeview Research.

Bachman F.P. 2010. *The Story of Inventions*. Airlington Heights, IL. Christian Liberty Press.

Barlow A. 2005. *The DVD revolution: movies, culture, and technology*. Westport, Conn., Praeger.

Baron R.A., Shane S.A. 2007. *Entrepreneurship: a process perspective*. Mason, OH. Thomson/South-Western. (online: dx.doi.org/10.1111/j.1540-6520.2011.00452.x)

Bilous R., et al. 2010. *Handbook of Diabetes*. Chichester, West Sussex – Hoboken, NJ. John Wiley and Sons. (online: dx.doi.org/10.1002/9781444391374.ch30)

Brzeziński T. (ed.) 1995. *Historia Medycyny*, wyd. II. Warszawa. PZWL.

Baskaran M. 2011. 'Environmental Isotope Geochemistry': Past, Present and Future [in:] M. Baskaran (ed.) *Handbook of Environmental Isotope Geochemistry*, Vol. 1. Berlin. Springer.

Beaudouin D. 2005. *Charles Beaudouin: Une histoire d'instruments scientifiques*. Les Ulis. EDP Sciences.

Betz F. 2011. *Managing Technological Innovation: Competitive Advantage from Change*. Hoboken, NJ. John Wiley and Sons.

Bunch B.H., Hellemans A. 2004. *The History of Science and Technology* (…). Boston. Houghton Mifflin.

Burgess J., et al. 1990. *Under the microscope: a hidden world revealed.* Cambridge – New York. Cambridge University Press.

Burns R.W. 2000. *John Logie Baird: television pioneer.* London. IET.

Carlisle R.P. 2004. *Scientific American inventions and discoveries: all the milestones in ingenuity* (…). Hoboken, NJ. John Wiley and Sons.

Chadwyck-Healey Ch. 2011. *The New Textual Technologies* [in:] S. Eliot, J. Rose, *A Companion to the History of the Book.* Hoboken. John Wiley and Sons. (online: dx.doi.org/10.1002/9780470690949.ch33)

Chant Ch. 2004. *Tanks.* Leicester. SB.

Cole D. J., et. al. 2003. *Encyclopedia of modern everyday inventions.* Westport Conn. – London. Greenwood Publishing Group.

Crawford W. 1988. *Current technologies in the library: an informal overview.* Boston, MA. G.K. Hall.

Daniel E. D., et al. 1999. *Magnetic recording: the first 100 years.* Piscataway, NJ. IEEE Press.

Davidson J.K. 2000. *Clinical diabetes mellitus: a problem-oriented approach.* New York. Thieme

Ellinger F. 2008. *Radio Frequency Integrated Circuits and Technologies.* Berlin. Springer.

Encyclopaedia Britannica: Gwinn R.P., et al. 1990. *The New Encyclopaedia Britannica,* 15th edition. Chicago – Auckland – Geneva – etc. Encyclopaedia Britannica (cited as Encyclopaedia Britannica).

Gazis D.C. 2002. *Traffic Theory.* Boston – Dordrecht – London. Kluwer.

Gleissner D. 2007. *Epilepsie – ein Eingriff in das Leben junger Menschen: mit Bezugnahme auf ein narratives Interview und daran anschließende Analyse.* München. GRIN Verlag.

Grant R.M. 2002. *Contemporary strategy analysis: concepts, techniques, applications.* Malden, MA – Oxford. Wiley-Blackwell

Gross L.P., et al. 2003. *Image ethics in the digital age.* Minneapolis, MN. University of Minnesota Press.

Gwinn R.P. et al. – see *Encyclopaedia Britannica.*

Hall C.W. 2008. *A biographical dictionary of people in engineering: from the earliest records until 2000.* West Lafayette, IN. Purdue University Press.

Herwig K.W. 2009. Introduction to the neutron [in:] I.S. Anderson, et al. (eds.) *Neutron Imaging and Applications: A Reference for the Imaging Community.* New York – London. Springer. pp. 3-12.

Hillman D., Gibbs D. 1999. *Century makers: one hundred clever things we take for granted which have changed our lives over the last one hundred years.* New York. Wellcome Rain.

Hounshell D.A., Smith J.K. 1988. *Science and Corporate Strategy: Du Pont R&D, 1902--1980.* Cambridge – New York – Oakleigh. Cambridge University Press.

Human Rights Watch. 2007. *Survey of Cluster Munition Policy and Practice.* [New York]. Human Rights Watch (Cited as: Human Rights Watch 2007)

Iliffe C.E. 1984. *An introduction to nuclear reactor theory.* [Greater] Manchester – Dover, NH. Manchester University Press.

Jaffuel D., et al. 2001. *Les allergies alimentaires* [in:] *Revue Française d'Allergologie et d'Immunologie Clinique,* Vol. 41, Issue 2. pp. 187-198.

Kilby J.S. 2000. Turning Potential into Realities: The Invention of the Integrated Circuit [at:] http://www.nobelprize.org/nobel_prizes/physics/laureates/2000/kilby-lecture.html (accessed 2014-01-23).

Khazan O. 2013. Douglas Engelbart, computer visionary and inventor of the mouse, dies at 88 [in:] *The Washington Post*, 2013-07-03, available online [at:] http://www.washington-post.com/business/douglas-engelbart-computer-visionary-and-inventor-of-the-mouse-dies-at-88/2013/07/03/1439b508-0264-11e2-9b24- ff730c7f6312_story.html (accessed 2014-08-28).

Klooster J.W. 2009. *Icons of invention: the makers of the modern world from Gutenberg to Gates*. Santa Barbara, CA. Greenwood Press.

Knoll M., Ruska E. 1932. Das Elektronenmikroskop [in:] *Zeitschrift für Physik A Hadrons and Nuclei*, Volume 78, Numbers 5-6 (1932). pp. 318-339.

Kragh H. 2002. *Quantum Generations: A History of Physics in the Twentieth Century*. Princeton, N.J.; Oxfordshire. Princeton University Press.

de Lacey L. 2007. *Australia's Greatest Inventions: From Boomerang to the Hills Hoist*. Wollombi, N.S.W. Exisle Publishing.

Landsteiner K. 1901. Über Agglutinationserscheinungen normalen menschlichen Blutes [in:] *Wiener klinische Wochenschrift*, Vol. 14. pp. 1132-1134.

L'Annunziata M.F. 2007. *Radioactivity: introduction and history*. Amsterdam. Elsevier.

Macnamara J. 2010. *The 21st century media (r)evolution: emergent communication practices*. New York. Peter Lang.

Malaga R.A. 2005. *Information systems technology*. Upper Saddle River, NJ. Pearson Prentice Hall.

Manovich L. 2002. *The language of new media*. Cambridge, MA – London. MIT Press.

Messadié G. 1995. *120 odkryć, które zmieniły świat; leksykon* (Les grandes dévouvertes de la science, tr. D. Bralewski). Łódź. Opus.

Michaelis A.R. 1962. *How nuclear energy was foretold* [in:] *New Scientist*, Vol. 13, No. 276, 1 March 1962. pp. 507-509.

Moan J.L., Smith Z.A. 2007. *Energy Use Worldwide: A Reference Handbook. Santa Barbara, CA. ABC-CLIO*.

Monk P.M.S. 2004. *Physical chemistry: understanding our chemical world*. Chichester – Hoboken, NJ. John Wiley and Sons.

Mutt V. 1982. Chemistry of the Gastrointestinal Hormones and Hormone-like Peptides and a Sketch of Their Physiology and Pharmacology [in:] H. Hopkins (ed.) *Vitamins and Hormones*, vol. 39. New York. Academic Press. pp. 231-427.

Pérez S., Samain D. 2010. Structure and Engineering of Celluloses [in:] D. Horton, *Advances in Carbohydrate Chemistry and Biochemistry*, vol. 64. Amsterdam – Boston – etc. Elsevier. pp. 25-116. (online: dx.doi.org/10.1016/S0065-2318(10)64003-6)

Perskiy C. 1901. Télévision au moyen de l'électricité [in:] M.E. Hospitalier, *Congrès international d'électricité (Paris, 18-25 août 1900)*. Paris. Gauthier-Villars. pp. 54-56.

Phan G. 2008. *Etude structurale du systeme d'efflux membranaire MexXY-OprM impliqué dans la resistance aux antibiotiques chez Pseudomonas aeruginosa*. These pour l'obtention du grade de docteur de l'Universite Paris Descartes, présentée et soutenue publiquement par Gilles Phan, le 12 juin 2008. Paris. Universite Paris Descartes.

Pirquet C.V. 1906. Allergie [in:] *Münchener medizinische Wochenschrift*, 53. München.

Poole H. et al. 2005. *The Internet: a historical encyclopedia*. Santa Barbara, CA. ABC-CLIO.

Reilly E.D. 2003. *Milestones in computer science and information technology*. Westport, Conn. Greenwood Publishing Group.

Rembieliński R., Kuźnicka B. 1987. *Historia Farmacji*. Warszawa. Państwowy Zakład Wydawnictw Lekarskich.

115

Rosenfeld L. 1982. *Origins of clinical chemistry: the evolution of protein analysis.* New York. Academic Press.

Rosenfeld L. 2002. Insulin: Discovery and Controversy [in:] *Clinical Chemistry 48:12.* pp. 2270-2288.

Roy A.A. 2001. *A history of the personal computer: the people and the technology.* London, Ont. Allan Publishing.

Safko L., Brake D.K. 2009. *The social media bible: tactics, tools, and strategies for business success.* Hoboken, NJ. John Wiley and Sons.

Schaeffler J. 2008. *Digital Signage: Software, Networks, Advertising, and Displays: A Primer for Understanding the Business.* Amsterdam – Boston. Focal Press.

Seeber G. 2003. *Satellite Geodesy.* Berlin. Walter de Gruyter.

Segrave K. 2004. *Lie detectors: a social history.* Jefferson, NC. McFarland.

Senning A. 2007. *Elsevier's Dictionary of Chemoetymology.* Amsterdam. Elsevier.

Shaw R. 2003. *Great Inventors and Inventions.* Carlton, South Vic. Curriculum Press.

Smith R.W. 2009. *Great Inventions & Inventors Spotlight on America.* Westminster, CA. Teacher Created Resources.

Soddy F. 1913. Radioactivity [in:] *Annual Reports on the Progress of Chemistry*, 1913. pp. 262-288.

Starling E.H. The Croonian lectures on the chemical correlation of the functions of the body [in:] *Lancet* 1905: 2. pp. 339-341.

Steed R.G. 2007. *Nuclear Power in Canada and Beyond.* Renfrew, Ont. General Store Publishing House.

Tarrant J. 2007. *Understanding Digital Cameras: Getting the Best Image from Capture to Output.* Oxford – Burlington, MA. Focal Press.

Tucker S.C., Roberts, P.M. (eds.) 2006. *World War One – Student Encyclopedia.* Santa Barbara – Denver – Oxford. ABC-CLIO.

Vickery H.B. 1950. The Origin of the Word Protein [in:] *Yale Journal of Biology and Medicine*, 1950 May; 22(5). pp. 387-393.

Volti R. 1999. *The Facts On File Encyclopedia of Science, Technology, and Society*, vol. 2. New York. Facts on File.

Wikipedia. *Computer file* [at:] http://en.wikipedia.org/wiki/Computer_file (accessed 2014-08-28).

Wikipedia. *Дастгохи майкровейв* [at:] http://tg.wikipedia.org/wiki/ %D0%94%D0%B0% D1%81%D1%82%D0%B3%D0%BE%D2%B3%D0%B8_%D0%BC%D0%B0%D0% B9%D0%BA%D1%80%D0%BE%D0%B2%D0%B5%D0%B9%D0%B2 (accessed 2014-07-12).

Williams D.B., Carter C.B. 2009. *Transmission Electron Microscopy: A Textbook for Materials Science.* New York – London. Springer.

Yasuda H. 2011. *Magneto Luminous Chemical Vapor Deposition.* Boca Raton, FL. CRC Press.

Yeo K.S., et al. 2010. *Intellectual Property for Integrated Circuits.* Fort Lauderdale, FL. J. Ross Publishing.

Yost J.R. 2005. *The Computer Industry.* Westport, Conn. Greenwood Publishing Group.

Zhelyazova B. 2005. *Elsevier's Dictionary of Automation Technics: In English, German, French and Russian.* Amsterdam. Elsevier.

Ziółkowska M. 1997. *Skąd my to mamy? : dzieje przedmiotów codziennego użytku.* Warszawa. Wydawnictwo Naukowo-Techniczne.

Zorei H. 2011. Barrasi-i ta'sir-i fanovari-i ittiloʔot va irtibotot bar bonk-i millatho-i Eron [in:] *Payom-i Donišgoh-i milli-i Tožikiston*. 11(75). Part II. Dušambe. Sino. pp. 103-106.

5.4 Linguistic materials

Abbāsi H. 1387-06-27 HŠ. Darvāze-i barā-ye gašt-o-gozār dar žahān-veb [at:] Khorasan-news (website) http://www.khorasannews.com/News.aspx?type=7&year=1387&month=6&day=27&id=3156400 (accessed 2014-08-15).

Abdulloyeva M. 2010-04-28. Šinosnoma-i čipdor [at:] http://paikon.wordpress.com/2010/04/28/%D1%88%D0%B8%D0%BD%D0%BE%D1%81%D0%BD%D0%BE-%D0%BC%D0%B0%D0%B8-%D1%87%D0%B8%D0%BF%D0%B4%D0%BE-%D1%80/ (accessed 2012-08-10).

Abdulov K. 2010-02-17. Karim Abdulov's website [at:] http://www.abdulov.tj/tj.html (accessed 2010-10-06).

Abdulvohid M. (date unknown). Sinamo-i tožik dar hol-i šukufoi [at:] http://iransharghi.com/engine/print.php?newsid=1784&news_page=1 (accessed 2011-10-09).

Abdurahmonova D. 2010-10-21. Harflar olami [at:] http://tashgiuu.zn.uz/46 (accessed 2011-09-14).

Abdurrahmon A. (date unknown). Az Misr či haqiqat-e kašf šud? In ham muʔžiza-e az Qurʔon [at:] http://www.wasatiyat.tj/component/content/article/2-2010-06-23-06-41-17/667-2011-02-11-13-19-56.html (accessed 2011-04-04).

Abrorova. 2009-09-30. *kmid.xml* (Tajik translation) [at:] http://opus.lingfil.uu.se/KDE4/xml/tg/messages/extragear-multimedia/kmid.xml.gz (accessed 2011-09-16).

Aftab (website). 1386-04-16 HŠ. Internet va veb [at:] http://www.aftabir.com/articles/view/computer_internet_infortmation_technology/internet_network/c14c1183813146_internet_web_p1.php/%D8%A7%DB%8C%D9%86%D8%A-A%D8%B1%D9%86%D8%AA-%D9%88-%D9%88%D8%A8 (accessed 2014-09-27).

Agroinform.tj (website). 2010. Bozi-i burdnok dar Agroinform.tj [at:] http://new.agroinform.tj/images/gazeta/newspaper_design_taj.pdf&sa=U&ei=6IuoTfzsHsPssgafos2dCA&ved=0CAwQFjAA&usg=AFQjCNFTkjRyabCWRnyGkleyXV98XhSF4Q (accessed 2011-04-05).

Agroinvestbonk. 2011-06-15. Agroinvestbonk (*advertisement*) [in:] *Čarx-i Gardun* No. 24 (771). p. 24.

Ahmadi M. 2011-06-09. Ošti-i oila va pušaymoni az taloq-i SMS-i [at:] *Radyo-i Ozodi* http://www.ozodi.tj/content/article/24230102.html (accessed 2011-08-08).

Al-Arabiyya (website). 2011-06-09. Google [gūgl] yaḥtafil bi-l-dikrā al-96 li-mīlād muhtaraʿ al-ǧītār al-kahrabāʾī [at:] http://www.alarabiya.net/articles/2011/06/09/152592.html (accessed 2011-06-09).

Amonatbonk (website). (date unknown). «Bankomat»-ho-i Amonatbonk meafzoyad [at:] http://www.amonatbonk.tj/news/1/385/ (accessed 2011-01-31).

Anžoman-e rāhnomāy-e xānwāda-ye Afǧān. (date unknown). Barnāma-ye āgāhi wa weqāya-ye eč-āy-wi (maraz-e eydz) [at:] http://www.afga.org.af/dari/stra-plan-HIV_AIDS.htm (accessed 2010-12-25, cited as ARXA acc. 2010-12-25).

Arabwebcafe (website). 2011-08-28. maqhā al-wīb al-ʕarabī [at:] http://www.arabwebcafe.com (accessed 2011-10-06).

ARXA – *see* Anǰoman-e rāhnomāy-e xānwāda-ye Afğān.

Asia-Plus (website). 2013-08-07. Kumita-i televizyon va radyošunavoni ba Kabiri baro-
-i paxš-i raddiya-i vaqt ǰudo namekunad [at:] http://news.tj/tj/news/kumitai-televizi
on-va-radioshunavon-ba-kabir-baroi-pakhshi-raddiya-va-t-udo-namekunad (accessed
2013-12-10).

Ašŭrov A. 2014-03-12. Ixtiro-i muhandis-i toǰik baro-i batareyaho-i oftobi [at:] Radyo-i
Ozodi http://www.ozodi.mobi/a/tjk-scientist-discovered-special-controller-for-solar-
batteries/25294591.html (accessed 2014-09-15).

Ašurova R. 2011-05-11. R. Ašurova dar bora-i bemori-i SPID (*broadcast*) [at:] *Radyo-i Ozodi* http://www.ozodi.tj/content/article/24098387.html (accessed 2011-06-08).

Atoyev A. (date unknown). Mohiya-i NOKA dar tahkim-i zabon-i Toǰiki [at:] http://www.
policy.hu/atoev/articles/FOSS4TjLanguage.pdf (accessed 2011-10-06).

ATU – *see* Azərbaycan Tibb Universiteti.

Axtar D. (date unknown). Šahr boyad šahrdor došta bošad na xoǰagidor [at:] Markazi Tadqiqot-i Žurnalisti http://mtjt.tj/index.php?NewID=159 (accessed 2011-03-28).

Ayubzod S. 2011-05-11. Pešnihod-i nerŭgoh-i hastai dar ivaz-i «Roğun» (*broadcast*)[at:] *Radyo-i Ozodi* http://www.ozodi.org/content/article/16799647.html (accessed 2011-07-10).

Ayubzod S. 2011-05-29. Detektor-i durŭğ [at:] http://aioubzod.wordpress.com/2011/05/29
/%D0%B4%D0%B5%D1%82%D0%B5%D0%BA%D1%82%D0%BE%D1%80%D0
%B8-%D0%B4%D1%83%D1%80%D3%AF%D2%93 (accessed 2014-07-08).

Ayubzod S. 2012-04-25. Video-e, ki šoyad girya-i myhoǰiron-ro orad [at:] *Radyo-i Ozodi*
http://www.ozodi.org/content/blog/24559174.html?s=1 (accessed 2012-11-11).

Azərbaycan Tibb Universiteti (website). (date unknown). Tibbi mikrobiologiya və immu-
nologiya kafedrası – Tədris proqramı [at:] http://www.amu.edu.az/az/cafedra/463/473
(accessed 2014-05-15, cited as ATU, acc. 2014-05-15).

Azizi Z., Golbān Moqaddam N. 1346 HŠ. Partow-e lāzer va mavāred-e este'māl-e ān dar
pezeški [in:] *Tehran University Medical Journal 24/8. 1346 HŠ.* pp. 653-668.

Barotov F., Gulxoǰa Š. 2010-12-01. Muškil-i bemori-i VIČ SPID (*broadcast*). [at:] *Radyo-i
Ozodi* http://www.ozodi.tj/content/article/2236144.html (accessed 2011-06-08).

Barzgar, M. 1390-07-04. Xorākpaz-e māykroveyv. [at:] oloumbuein blog http://oloumbuein.
blogfa.com/post-17.aspx (accessed 2015-01-28).

Bakhtarnews (website). 2011-04-11. Woqu?-e zelzela dar J̌āpān [at:] http://www.bakhtar-
news.com.af/da/index.php?news=20533 (accessed 2011-07-10, cited Bakhtarnews acc.
2011-04-11).

Bānk-e Sāderāt-e Yazd (website). 1388-02-05 HŠ. Dastgāh-e xodpardāz [at:] http://bsy.ir/
persian/index.php?option=com_content&task=view&id=137&itemid=84 (accessed
2011-07-20).

Bānk-e Teǰārat (website). (date unknown). Servis-e payām-e kutāh-e teǰārat [at:] http://www.
tejaratbank.ir/portal/default.aspx?tabid=414 (accessed 2011-08-09, cited as: Bānk-e
teǰārat, acc. 2011-08-09).

Baxtiyor. 2009-10-06. Vebblog-i Baxtiyor: Šinosoi [at:] http://bakhtiyor78.blogspot.
com/2009/10/blog-post.html (accessed 2011-10-06).

BBC Pashto (website). 2010-12-01. Afğānistān: də eyḍz… [at:] http://www.bbc.co.uk/pash-
to/afghanistan/2010/12/101201_aids-afghanistan.shtml (accessed 2010-12-25).

BBC Persian (website). 2004-05-17. Barnoma-i tafrehi-i toǰiki [at:] http://www.bbc.co.uk/
persian/tajikistan/story/2004/05/040517_dr_tajentertainment_cyr.shtml (accessed
2005-12-02).

BBC Persian (website). 2008-06-05. Manʔ-i istifoda az mobayl ba hangom-i ronandagi [at:] http://www.bbc.co.uk/persian/tajikistan/story/2008/06/printable/080605_d_rm_mobile_ cyr.shtml (accessed 2012-11-11).

BBC Persian (website). 2009-02-05. Duzdon-i daryoi kišti-i Ukroyin-ro tark kardand [at:] http://www.bbc.co.uk/tajik/news/2009/02/090205_ez_somalia_pirace.shtml (accessed 2011-01-09).

BBC Persian (website). 2009-05-14. Zanon ba bemoriho-i ufuni muqovimtarand [at:] http://www.bbc.co.uk/tajik/news/2009/05/090514_rm_women_immune.shtml (accessed 2011-03-24).

BBC Persian (website). 2009-07-16. ʒaʔbaho-i siyoh-i havopaymo-i eroni 'oseb didaast' [at:] http://www.bbc.co.uk/tajik/news/2009/07/090716_rm_planecrash_caspian.shtml (accessed 2011-09-20)

BBC Persian (website). 2009-07-17. Navoz Šarif tabraʔa šud [at:] http://www.bbc.co.uk/ tajik/news/2009/07/090717_ez_pakistan_nawaz.shtml (accessed 2011-01-06).

BBC Persian (website). 2009-09-15. Afsurdagi umr-i bemoron-i saratoni-ro kam karda-ast [at:] http://www.bbc.co.uk/tajik/news/2009/09/090915_rm_depression_cancer.shtml (accessed 2011-03-24).

BBC Persian (website). 2009-09-15. Haft ganʒ: ʒozibaho-i bartar-i turisti [at:] http://www. bbc.co.uk/tajik/news/2009/09/090915_if_wonder.shtml (accessed 2011-08-08, cited as BBC Persian, 2009-09-15b).

BBC Persian (website). 2009-11-10. Afzoiš-i ibtilo ba VIČ az roh-i žinsi [at:] http://www.bbc. co.uk/tajik/news/2009/11/091110_ez-sq-hiv-transmission.shtml (accessed 2011-06-11).

BBC Persian (website). 2010-01-07. Telefon-i hamroh mumkin ast az alzoymer pešgiri ku-nad [at:] http://www.bbc.co.uk/tajik/news/2010/01/100107_rm_cell-phones-alzheimers. shtml (accessed 2011-02-10).

BBC Persian (website). 2010-04-21. Zarbu. „www"lažal baro-i taxliya-i ižbori-i yak xob-goh [at:] http://www.bbc.co.uk/tajik/news/2010/04/100420_sq_hostel.shtml (accessed 2010-07-09).

BBC Persian (website). 2010-12-01. Afzoiš-i ibtilo ba EYDZ dar Xatlon [at:] http://www. bbc.co.uk/tajik/news/2010/12/101201_ea_ak_hiv_eids.shtml (accessed 2011-06-11)

BBC Persian 2012-11-14. Tibb-i gurůh-i xuni dar Žopun; gurůh-i xuni-i šumo či megůyad? [at:] http://www.bbc.co.uk/tajik/news/2012/11/121112_rm_japan_blood_type_fever. shtml (accessed 2014-05-14).

BMT – *see* Bonk-i Milli-i Toʒikiston

Bobiyev Ḡ.M., et al. 2007. *Ximiya. Kitob-i darsi baro-i sinf-i 11*. Dušambe. ʒMM XEROXland.

Boinazarov B., et al. 2007. Sanʔat va Mehnat. Kitob-i darsi baro-i sinf-i 3. Dušambe. Sar-parast.

Boinazarov B., et al. 2007. Sanʔat va Mehnat. Kitob-i darsi baro-i sinf-i 4. Dušambe. Sarpar-ast (cited as Boinazarov, et al. 2007b).

Bonk-i Milli-i Toʒikiston. 2011-01-12. Tavsif-i hisobho-i tavozuni [at:] http://nbt.tj/files/ pl_schetov/tavsif_plsch_kb_tj.pdf&sa=U&ei=5TeKTZ_xFJGBhQfZsqy9Dg&ved=0C BYQFjAEOAo&usg=AFQjCNH149gQGnn1s5FQQDA_Sz9R-xOs4w (accessed 2011-03-23, cited as BMT 2011-01-12).

Bonk-i Rušd-i Toʒikiston (website). (date unknown). Namoyandagiho va filialho [at:] http:// www.brt.tj/main1280.php?t=3&i=5&l=2 (accessed 2012-01-08, cited as BRT acc. 2012-01-08).

119

Boqizoda A. (date unknown). Eʔǰoz-i ilmi-i Qurʔon [at:] http://wasatiyat.tj/2010-10-19-12-23-12/274-2010-09-14-03-15-42.html (accessed 2011-03-25, cited as Boqizoda acc. 2011-03-25).

Botanix (website). 2009-06-06. Šinonidan-i piyozgulho va niholho [at:] http://www.botanix. kpr.eu/tg/index.php?text=4 (accessed 2011-08-11).

BRT – *see* Bonk-i Rušd-i Toǰikiston.

Cahan oğlu Əliyev S., et al. (date unknown). Qazanılmış immun çatışmazlığı sindromu [at:] http://kayzen.az/blog/immun/871/qazan%C4%B1lm%C4%B1%C5%9F- immun-%C3 %A7at%C4%B1%C5%9Fmazl%C4%B1%C4%9F%C4%B1-sindromu.html (accessed 2011-09-21, cited as Cahan oğlu Əliyev, acc. 2011-09-21).

Central Asian Voices (website). (date unknown). Qoidaho-i Umumi-i kor bo kompyuter [at:] http://www.centralasianvoices.tj/index.php?lng=tj&id=251&PHPSESSID=3140cfd3a 438b2cc32c842c318fe43f7 (accessed 2011-03-20, cited as Central Asian Voices, acc. 2011-03-20).

Central Asian Voices (website). (date unknown). Tafsir-i somonaho-i ba biznes-i fosilavi-i taʔlimi baxšida-šuda [at:] http://centralasianvoices.tj/index.php? lng=tj&id=286&PHPS ESSID=509557e0b7438023a0694064c5014601 (accessed 2011-03-22, cited as Central Asian Voices, acc. 2011-03-22).

Čarx-i Gardun. 2011-01-20. Terrorist dar Sulton-i Kabir [at:] http://www.gazeta.tj/chg/1355-terrorist-dar-sultoni-kabir.html (accessed 2011-09-11).

Čarx-i Gardun. 2011-08-04. Ǯaninšinosi-i Qurʔon [at:] http://gazeta.tj/chg/2839-1206anin-shinosii-1178uron.html (accessed 2014-09-12).

Čarx-i Gardun. 2012-06-08. SMS – payom-i oxirin ba Xolmůʔmin Safarov? [at:] http://gazeta. tj/chg/5032-sms-payomi-oxirin-ba-xolm1263min-safarov.html (accessed 2012-12-30).

Čarx-i Gardun. 2012-11-15. Kompyuter – boziča-i xatarnok [at:] http://gazeta.tj/chg/6074-kompyuter-bozichai-xatarnok.html (accessed 2013-02-17).

Dānešgāh-e Tarbiyyat-e Modarres. 1391-09-06. Fāylhā-ye xod-rā be šive-ye qeyr-e qābel-e bāzyābi pāk konid [at:] http://www.modares.ac.ir/research/TmuIT/edu/edu6 (accessed 2013-02-17, cited as DTM).

Dariush. 2008-02-24 7:06 am. *Follower*-i aržmand – *a comment posted under the article:* Rustam. 2008-02-11. Rahbar-i rahgumzada yo millat-i vayronšuda? [at:] http://toj var.wordpress.com/2008/02/21/%D1%80%D0%B0%D1%85%D0%B1%D0%B0%D1 %80%D0%B8-%D1%80%D0%B0%D1%85%D0%B3%D1%83%D0%BC%D0%B7% D0%B0%D0%B4%D0%B0-%D1%91-%D0%BC%D0%B8%D0%BB%D0%BB%D0 %B0%D1%82%D0%B8- %D0%B2%D0%B0%D0%B9%D1%80%D0%BE%D0%BD %D1%88/ (accessed 2014-06-22, cited as Dariush 2008-02-24).

DAT – *see* Donišgoh-i Agrari-i Toǰikiston.

Donišgoh-i Agrari-i Toǰikiston (website). (2011-03-23). Ilm [at:] http://www.tajagroun.tj/ index.php?option=com_content&task=view&id=53&Itemid=35 (accessed 2011-03-23, cited as DAT 2011-03-23).

Deutsche Welle [Arabic] (website). 2008-04-30. 15 ʿāman ʿalā wilādat al-šabaka al-ʿankabūtiyya al-ʿālamiyya [at:] http://www.dw.de/15-%D8%B9%D8%A7%D9%85%D8%A7%D9% 8B-%D8%B9%D9%84%D9%89-%D9%88%D9%84%D8%A7%D8%AF%D8%A9-% D8%A7%D9%84%D8%B4%D8%A8%D9%83%D8%A9-%D8%A7%D9%84%D8% B9%D9%86%D9%83%D8%A8%D9%88%D8%AA%D9%8A%D8%A9-%D8%A7% D9%84%D8%B9%D8%A7%D9%84%D9%85%D9%8A%D8%A9-www/a-3302413 (accessed 2011-10-06).

DMT – *see* Donišgoh-i Milli-i Tožikiston.

Donišgoh-i Milli-i Tožikiston (website). 2010-11-04. Informatika [at:] http://www.tgnu.tarena. tj/savtest/mekh-mat/informatika/2010_11_04_1_727_008_55.rtf (accessed 2011-07-03, cited as DMT 2010-11-04).

Donišgoh-i Milli-i Tožikiston (website). 2010-06-02. Informatika – Islomov [at:] http://www.tgnu.tarena.tj/savtest/testhoikrediti/Shahrak/02.06.2010 (accessed 2011-07-03, cited as DMT 2010-06-02).

Donišgoh-i Milli-i Tožikiston (website). 2010-04-15. Mažmŭ-i savolho-i testi (sanžiši) az fann-i tahlil-i sistemavi [at:] http://www.tgnu.tarena.tj/savtest/huquq/huquqi%20 sudi%20va%20nazprokurori/12.doc (accessed 2010-09-22, cited as DMT 2010-04-15).

Donišgoh-i Milli-i Tožikiston (website). 2010-04-08. Savolnoma-i informatika [at:] http:// www.tgnu.tarena.tj/savtest/iqtisod%2520va%2520idora/Kaf%2520tijorat/1%25200719 .doc&sa=U&ei=Hj6GTeGvB4OxhQe058m4BA&ved=0CA4QFjAB&usg=AFQ jCNHOqObl8ZgS6HCkWUQJ_jUiOoF1QQ (accessed 2011-03-20, cited as DMT 2010-04-08).

Donišgoh-i Milli-i Tožikiston (website). 2010-05-14. Texnologiya [at:] http://www.tgnu. tarena.tj/savtest/moliyaviyou%20iqtisodi/kaf%20iqtisod%20va%20idorai%20sayohi/% d0%a2%d0%b5%d1%85%d0%bd%d0%be%d0%bb%d0%be%d0%b3%d0%b8%d1% 8f.doc (accessed 2011-07-09, cited as DMT 2010-05-14).

Donišgoh-i Milli-i Tožikiston (website). 2010-06-07. Informatika, Rustamova, X 1k-KKKF [at:] http://www.tgnu.tarena.tj/savtest/testhoi%20krediti/Markaz/07.06.2010/%d0%98 %d0%bd%d1%84%d0%be%d1%80%d0%bc%d0%b0%d1%82%d0%b8%d0%ba%d 0%b0%20%d0%a0%d1%83%d1%81%d1%82%d0%b0%d0%bc%d0%be%d0%b2% d0%b0%20%d0%a5%201%d0%ba-%d0%9a%d0%9a%d0%9a%d0%a4.rtf (accessed 2011-05-07, cited as: DMT 2010-06-07).

Donišgoh-i Milli-i Tožikiston (website). 2011-[04-22]. Fakultet-i fizika; Kafedra-i Mošinho-i hisobbaror, sistemaho va šabakaho (…) [at:] http://www.tgnu.tarena.tj/savtest2011/fizi-ka/moshinhoi%20hisob/k2/1.doc (accessed 2011-09-18, cited as DMT 2011-[04-22]).

Donišgoh-i Milli-i Tožikiston (website). [2011-12-02]. [6.doc] [at:] http://www.tnu.tj/ savtest2011-2012/Fizika/kaf%20MH%20va%20SSh/kursi3/6.doc (accessed 2012-08-11, cited as DMT [2011-12-02]).

Dostiyev A. 2009-11-19. Tožikiston tanho az Dušambe iborat nest yo davlat-ro boyad davlat bisozad. [at:] http://millat.tj/ijtimo/1585-tojikiston_tanho_az__394.html (accessed 2011-05-17).

DP-News (website). 2010-10-29. ḡītār kahrabāʔī… [at:] http://www.dp-news.com/pages/de tail.aspx?articleid=60854 (accessed 2011-10-03)

DTM see Dānešgāh-e Tarbiyyat-e Modarres.

Elshami (website). [2010-11-12]. World Wide Web [at:] http://www.elshami.com/Terms/W/ World%20Wide%20Web.htm (accessed 2011-10-06).

Faraj.tj (website). 2014-05-29. Gufted? Megŭyand [at:] http://faraj.tj/paradox/5406-gufted-meg1263yand-22-2014.html (accessed 2014-09-11).

Farhodi Ӡ. 2003-02-25. Vazʔ-i sarbozon-i yak bataliyon-i tanki dar nohiya-i Danḡara [at:] Radyo-i Ozodi (website) http://www.ozodi.org/content/article/1122818.html (accessed 2011-05-29).

Farmon F. 2013-05-25. Saida kelin-i Ŭzbekho mešavad? [at:] faraj.tj (website) http://faraj. tj/culture/2274-saidai-siro1207iddin-kelini-1263zbek1203o-meshavad.html (accessed 2014-09-11).

Farzin. 2008-08-01. Naslho-i MEH [at:] http://www.farzin.tj/Matn22.html (accessed 2011-01-24).

Firůz M., Šarif M. (date unknown). Xatar-i VIČ/SPID dar šahr-u dahot yakson ast [at:] Jumhuriyat (website) http://www.jumhuriyat.tj/index.php?art_id=481 (accessed 2011-06-08).

Habibullayev P.K., et al. 2010. *Fizika. Kitob-i darsi baro-i donišomůzon-i sinfho-i 9-um-i maktabho-i taʔlim-i umumi-i miyona.* Toškand. Xona-i eʒodi-i tabʔ-u našr-i ba nom-i Ḡafur Ḡulom.

Hamad S. 2009-06-29. Anʒom-i Anvaroššo az silsila-i Bosmači ki bud? Vatanxoh yo xoin? [at:] *Millat* (website) http://millat.tj/tarih/1388-anjomi_anvarposhsho__668.html (accessed 2011-01-01).

Hamidov U. (date unknown). Baland bardoštan-i maʔrifat-i huquqi-i navrason- omil-i pešgiri namudan-i ʒinoyatkori dar bayn-i onho [at:] *Maktab* (website). http://www.maktab.tj/reports.php (accessed 2011-05-17).

Hizb-i Nahzat-i Islomi-i Toʒikiston (website). 2011-05-13. Q. Rasulzoda baro-i taʔmin-i fazo-i ittilooti 1 mln somoni ʒudo kard [at:] http://nahzat.tj/k/1/t/2859--1-.html (accessed 2011-05-15, cited as Hizb-i Nahzat, 2011-05-13).

Hizb-i Nahzat-i Islomi-i Toʒikiston (website). 2011-12-20. Extimol-i paydoiš-i saraton- -i maḡzi az istifoda-i telefon-i dasti [at:] http://www.nahzat.tj/akhbor/item/4051 -%D1%8D%D2%B3%D1%82%D0%B8%D0%BC%D0%BE%D0%BB%D0%B8 -%D0%BF%D0%B0%D0%B9%D0%B4%D0%BE%D0%B8%D1%88%D0%B8 -%D1%81%D0%B0%D1%80%D0%B0%D1%82%D0%BE%D0%BD%D0%B8 -%D0%BC%D0%B0%D2%93%D0%B7%D3%A3-%D0%B0%D0%B7-%D0%B 8%D1%81%D1%82%D0%B8%D1%84%D0%BE%D0%B4%D0%B0%D0%B8 -%D1%82%D0%B5%D0%BB%D0%B5%D1%84%D0%BE%D0%BD%D0%B8- %D0%B4%D0%B0%D1%81%D1%82%D3%A3 (accessed 2012-11-11, cited as Hizb-i Nahzat 2011-12-20).

Institute for Professional Development – GBAO (website). (date unknown). Barnomaho-i xidmati va šabakaho-i kompyuteri [at:] http://www.ipd-gbao.org/offline/2005/8/1129024658/ (accessed 2011-03-23, cited as: IPD GBAO acc. 2011-03-23).

ICBL. 2009. International Campaign to Ban Landmines Press Release, 7 July 2009. ICBL.

IQRA (website). 2008-12-17. Donišomuzon-i Ironi payom-i hamdil-i baro-i kudakon-i Ḡazza irsol mekunand [at:] http://www.iqna.ir/tj/news_detail.php? ProdID=335063 (accessed 2011-08-08).

Islomov S., et al. 2005. Kasb-u Hunar (Texnologiya). Kitob-i darsi baro-i sinf-i 6 (baro-i pisarho). Dušambe.

Istad A. 2012-02-25. Mard-i ozodandeš [at:] http://gerbisherdor.blogspot.com/2012/02/blog-post_25.html (accessed 2012-08-09).

Ʒonmard D. 2010-01-22. Vusʔat-i farogiri-i paxš-i barnomaho-i televizyoni [at:] Khovar (website) http://www.khovar.tj/taj/index.php?option=com_content&task=view&id=113 12&Itemid=&bsb_midx=10 (accessed 2011-03-27).

Jumhuriyat (website). (date unknown). Du růydod-i farahbaxš [at:] http://jumhuriyat.tj/index.php?art_id=536 (accessed 2011-03-27, cited as Jumhuriyat acc. 2011-03-27).

Jumhuriyat (website). (date unknown). Lui Paster [at:] http://jumhuriyat.tj/index.php?art_id=674 (accessed 2011-04-04, cited as Jumhuriyat acc. 2011-04-04).

Jumhuriyat (website). (date unknown). Umedvoriho-i kišvarho-i 5+1 ba guftugů bo Eron [at:] http://www.jumhuriyat.tj/index.php?art_id=3046 (accessed 2011-07-09, cited as Jumhuriyat acc. 2011-07-09).

Jumhuriyat (website). (date unknown). Xabarho [at:] http://www.jumhuriyat.tj/index. php?art_id=10284 (accessed 2013-01-29, cited as Jumhuriyat acc. 2013-01-29).

Jumhuriyat (website). (date unknown). Qonun-i Ǯumhuri-i Toǯikiston Oid ba void namudan-i taḡyir-u ilovaho ba Qonun-i Ǯumhuri-i Toǯikiston "Dar borai muqovimat bo virus-i norasoi-i masuniyat-i odam va bemori-i norasoi-i muhassal-i masuniyat [at:] http:// www.jumhuriyat.tj/index.php?art_id=1064 (accessed 2014-09-07, cited as Jumhuriyat acc. 2014-09-07).

Jumhuriyat (website). (date unknown). Emomali Rahmon ba mahalho-i az ofat-i tabiat zarardida-i viloyat-i Xatlon safar kard [at:] http://www.jumhuriyat.tj/index.php?art_id=4930 (accessed 2014-09-10).

Ǯŭrayev S. (date unknown). Advokat eʔtiroz mekunad [at:] *Markaz-i Tadqiqot-i Žurnalisti* (website) http://mtjt.tj/index.php?NewID=163 (accessed 2011-03-27).

Kabirov Š., Ayubova M.B. 2009. Zabon-i toǯiki (dastur-i taʔlimi baro-i donišǯuyon-i fakultet-i byologiya [at:] http://www.tgnu.tarena.tj/savtest/testhoi %20umumidonishgohi/zaboni%20tojiki/zaboni%20modari%20baroi%20biologho.doc (accessed 2011-07-30).

Karim N. [2011]-05-17. Čorabini baxšida ba yodbud-i favtidagon az norasoi-i masuniyat-i odam [at:] http://tojnews.org/taj/index.php?option=com_content&task=view&id=11412 (accessed 2011-06-08).

Karim O. 2011-01-26. SMS-payomak vasila-i fasl [at:] http://orzuikarim.wordpress.com/ page/3/ (accessed 2011-08-08).

Karimzoda B. 2010. Kartoška – non-i duyum [in:] *Fermer* No. 3, noyabr-i sol-i 2010, pp. 23-27.

Khovar (website). 2008-08-21. Faʔoliyat-i širkatho-i aloqa-i telefon-i mobili muvaqqatan bozdošta šud [at:] http://khovar.tj/archive/173-faoliyati-shirkat1202oi- alo1178ai-telefoni-mobil1250-muva11781178atan-bozdoshta-shud.html (accessed 2012-11-11).

Khovar (website). [2010]-04-09. Bo sistema-i nav-i «SMS – maʔlumot» payvast šud [at:] http://khovar.tj/archive/12728-bo-sistemai-navi-sms-malumot-payvast-shud.html (accessed 2011-08-08).

Khovar (website). [2011]-07-21. Ḡoratgaron hanŭz dastgir našdaand [at:] http://khovar. tj/archive/7838-1170oratgaron-1202an1262z-dastgir-nashudaand.html (accessed 2011-09-11).

Killid Group, the (website). 1389-03-06 HŠ. Se māhvāra-ye ǯāpāni dar fazā nāpadid šoda ast [at:] http://www.tkg.af/dari/world-news/world-news/1757-%D8%B3%D9%87-%D9%85% D8%A7%D9%87%D9%88%D8%A7%D8%B1%D9%87-%D8%AC%D8%A7%D9% BE%D8%A7%D9%86%D9%8A-%D8%AF%D8%B1-%D9%81%D8%B6%D8%A7- %D9%86%D8%A7%D9%BE%D8%AF%DB%8C%D8%AF-%D8%B4%D8%AF% D9%87-%D8%A7%D8%B3%D8%AA (accessed 2010-11-20).

Killid Group, the (website). 1389-09-10 HŠ. Negarāni az afzāyeš-e bimāri-ye eydz dar kešwar [at:] http://tkg.af/dari/report/health/4061-%D9%86%DA%AF%D8%B1%D8% A7%D9%86%DB%8C-%D8%A7%D8%B2-%D8%A7%D9%81%D8%B2%D8%A7 %DB%8C%D8%B4-%D8%A8%DB%8C%D9%85%D8%A7%D8x%B1%DB%8C-% D8%A7%DB%8C%D8%AF%D8%B2-%D8%AF%D8%B1-%DA%A9%D8%B4%D 9%88%D8%B1 (accessed 2010-12-25).

Kimyo-i saodat (website). 2012-01-05. Čaro infiǯor-i buzurg yak nazariya-i ilmi namebošad? [at:] http://www.kemyaesaadat.com/2/index.php?option=com_content&view=article&i d=2261:1&catid=19:2008-12-09-19-20-33 (accessed 2012-01-05).

Komilov F.S., Šarapov D.S. 2003. Texnologiyaho-i informatsyoni. Kitob-i darsi baro-i xonandagon-i sinfho-i 8-9. Dušambe.

Komilov F.S., Šarapov D.S. 2005. Asosho-i Tehnika-i Kompyuteri. Kitob-i darsi baro-i xonandagon-i sinf-i 10. Dušambe. Matbuot

Kumita-i Andozi Nazd-i Hukumat-i Žumhuri-i Tožikiston. 2011-04-04. Davr-i nihoni-i ozmun-i žumhuriyavi-i «Tarbiya-i farhangi andozsupori-i nasl-i navras» [at:] http://an doz.tj/tg/node/315 (accessed 2011-04-15).

Kuvatov M. 2010-11-29. Translation of kdelibs4.po to Uzbek [at:] http://websvn.kde.org/ branches/stable/l10n-kde4/uz/messages/kdelibs/kdelibs4.po?revision=1207623&view= markup (accessed 2010-12-25).

LG. 2008. LPC53 User's Guide (LPC53_ENG-XO ARUSLLK_RUS). LG.

Lutfulloyev M. 2002. *Zabon-i modari, kitob-i darsi baro-i sinf-i 1*. Dušambe. Sarparast.

Lutfulloyev M., et al. 2007. *Alifbo. Kitob-i darsi baro-i sinf-i yakum*. Dušambe. Sarparast.

Mahkamova G. (date unknown). Duxtaron xondan namexohand yo onho-ro namemonand? [at:] Haqiqat-i Suğd (website) http://www.hakikati-sugd.tj/index.php/critics/1375-dukhtaron-khondan-namekhoand-jo-onoro-namemonand (accessed 2014-09-10; cited as Mahkamova, acc. 2014-09-10).

Mahmadbekova P. 2011-01-14. Boz šaš bankomat dar xizmat-i mizožon [at:] *Xovar* http://www.khovar.tj/taj/index.php? option=com_content&task=view&id=17007&Itemid=13 (accessed 2011-01-30).

Mažidov H., Nozimov O. 2006. *Fizika. Kitob-i darsi baro-i sinf-i 9*. [Dušambe]. Izdatel'stvo Učebnaja Literatura.

Mažlis-i Namoyandagon-i Mažlis-i Oli-i žumhuri-i Tožikiston. 2001-05-10. *Qaror-i Mažlis-i Namoyandagon-i Mažlis-i Oli-i Žumhuri-i Tožikiston «Dar bora-i sifat va bexatari-i mahsulot-i xůkvori» (...)* Dušambe (cited as Mažlis-i Namoyandagon 2001-05-10).

Mažlis-i Oli-i žumhuri-i Tožikiston. 2003-12-08. *Qonun-i Žumhuri-i Tožikiston dar bora-i baytori*. Dušambe (cited as Mažlis-i Oli 2003-12-08).

Mažlis-i Oli-i žumhuri-i Tožikiston. 2004-12-09. *Qonun-i Žumhuri-i Tožikiston dar bora-i tibb-i xalqi*. Dušambe (cited as Mažlis-i Oli 2004-12-09).

Mažlis-i Oli-i žumhuri-i Tožikiston. 2004. *Qonun-i Žumhuri-i Tožikiston dar bora-i ixtiroℇ* Dušambe (cited as Mažlis-i Oli 2004b).

Mažlis-i Oli-i žumhuri-i Tožikiston. 2005-03-01. *Qonun-i Žumhuri-i Tožikiston dar bora-i xadamot-i sadamavi (...)* Dušambe (cited as Mažlis-i Oli 2005-03-01).

Mažlis-i Oli-i žumhuri-i Tožikiston. 2009-04-01. *Qonun-i Žumhuri-i Tožikiston dar bora--i himoya-i tibbi-yu ižtimoi-i šahrvandon-i mubtalo-i diabet-i qand*. Dušambe (cited as Mažlis-i Oli 2009-04-01).

Mažlis-i Oli-i žumhuri-i Tožikiston. 2009-11-19. *Qonun-i Žumhuri-i Tožikiston dar bora-i qabul va mavrid-i amal qaror dodan-i Kodeks-i murofiavi žinoyati-i Žumhuri-i Tožikiston*. Dušambe (cited as Mažlis-i Oli 2009-11-19).

Mažlis-i Oli-i žumhuri-i Tožikiston. 2010-12-08. *Qonun-i Žumhuri-i Tožikiston dar bora-i baytori*. Dušambe (cited as Mažlis-i Oli 2010-12-08).

Mažlis-i Oli-i žumhuri-i Tožikiston. 2010-12-16. *Qonun-i Žumhuri-i Tožikiston dar bora-i himoya-i ižtimoi-i maℇyubon dar Žumhuri-i Tožikiston*. Dušambe (cited as Mažlis-i Oli 2010-12-16).

Malikov Q. 2009. Durnamo-i hamkoriho-i Eron va Sozmon-i Hamkori-i Šanxay [at:] *Iransharghi* http://www.iransharghi.com/engine/print.php?newsid=2644&news_page=1 (accessed 2011-03-27).

Maqomot-i Ižroiya-i Hokimiyat-i Davlati-i Viloyat-i Suğd. 2010-12-10. Iftitoh-i markaz-i ta-šxisi-yu tabobati-i lazeri dar Xužand. [at:] http://www.sugd.tj/index.php?option=comco

ntent&view=article&id=1585:2010-12-10-10-39-16&catid=19:2008-11-19-05-05-59& Itemid=13 (accessed 2011-03-25, cited as Viloyat-i Suğd).

Maqomot-i Iğroiya-i Hokimiyat-i Davlati-i Viloyat-i Suğd. 2011-03-30. Namoyanda-i Viloyat-i Suğd ğolib-i ozmun-i «Tarbiya-i farhangi andozsupori-i nasl-i navras» [at:] http://www.sugd.tj/index.php?option=com_content&view=article&id=1913:-----l----r& catid=19:2008-11-19-05-05-59&Itemid=13 (accessed 2011-04-15, cited as Viloyat-i Suğd 2011-03-30).

Markaz-i Milli-i Patent-u Ittiloot. 2011-07-21. *Navid-i Patenti. Xabarnoma-i rasmi*. No. 2 (62). Dušambe (cited as MMPI 2011-07-21).

Markaz-i Milli-i Patent-u Ittiloot. 2011-10-24. *Navid-i Patenti. Xabarnoma-i rasmi*. No. 65. Dušambe (cited as MMPI 2011-10-24).

Markaz-i Milli-i Patent-u Ittiloot. 2011-11-16. *Navid-i Patenti. Xabarnoma-i rasmi*. No. 66. Dušambe (cited as MMPI 2011-11-16).

Markaz-i Milli-i Patent-u Ittiloot. 2012-01-25. *Navid-i Patenti. Xabarnoma-i rasmi*. No. 68. Dušambe (cited as MMPI 2012-01-25).

Matrix-tv (website). (date unknown). Tugma-i ilovagi-i Win baro-i či darkor ast? 22 masli-hat-i mufid [at:] http://matrix-tv.tj/tugmai-ilovagii-win-baroi-chjj-darkor-ast-22-masli-hati-mufid/ (accessed 2014-09-12).

Media forum (website). 2011-01-18. Əhmədinejad: "Hey oturub desinlər ki, İran atom bombası istehsal etmək niyyətindədir" [at:] http://www.mediaforum.az/az/2011/01/18/%C6%8 FHM%C6%8FD%C4%B0NEJAD-HEY-OTURUB-DES%C4%B0NL%C6%8FR-K %C4%B0-%C4%B0RAN-ATOM-BOMBASI-044515762c03.html#.U6H0F2Yiedk (accessed 2011-09-21).

Millat (website). 2008-11-08. Noyabr-i xunin-i Dušambe-šahr [at:] http://millat.tj/ijtimo/950-noyabri_khunini_dush_315.html (accessed 2010-12-10).

Millat (website). 2009-02-04. Noma-i sipos ba xonandagon [at:] http://millat.tj/millatvais lom/1123-nomai_sipos_ba_khona_354.html (accessed 2009-02-04).

Mirbozxonova Ӡ. 2014-05-16. Ӡarima-i "čini" [at:] faraj.tj (website) http://faraj.tj/opinion/ 5306-1206arimai-chin1251.html (accessed 2014-07-02).

Mirzob S. 2010-12-03. Sarvaron-u merosbaron-i SSSR [at:] *Khatlonpress* (website) http:// www.khatlonpress.tj/index.php? option=com_content&task=view&id=540&Itemid=15 (accessed 2011-09-18).

MMPI – *see* Markaz-i Milli-i Patent-u Ittiloot.

Muhabbat va Oila (website). 2010-03-05. Obama IMA-ro kušt. [at:] http://muhabbatvaoila.tj/ simoi-shinohta/56-obama-ima-ro-kusht.html (accessed 2010-08-23).

Muhabbat va oila (website). 2010-07-01. Xomušak ki-ro megazad? [at:] http://muhabbatvao-ila.tj/tib/134-xomshak-kiro-megazad.html (accessed 2010-09-01).

Muhabbat va oila (website). 2010-12-16. Gribok dar uzvho-i žinsi [at:] http://muhabbatvao-ila.tj/tib/287-gribok-dar-uzv1202oi-1206ins1250.html (accessed 2011-03-22).

Muhammad F. 2011-01-20. Puštiboni-i Rusiya az taškil-i davlat-i Falastin bo poytaxt-i on [at:] *Tojnews* (website) http://tojnews.org/taj/index.php?option=com_content&task=vie w&id=10821&Itemid=3%20class=&bsb_midx=1 (accessed 2011-07-09).

Muhammad F. 2011-03-16. Parokanda šudan-i tazohurot-i Manama az žonib-i nerůho-i in-tizomi [at:] *Tojnews* (website) http://tojnews.org/taj/index.php?option=com_content &task=view&id=12542%20class=&bsb_midx=3 (accessed 2011-05-29).

Muhammadražab M. 2012-08-31. Sadho tifl-i šunavoi-aš zaʔif ba markaz-i maxsus niyoz dorad [at:] Radyo-i Ozodi http://www.ozodi.mobi/a/audiology-center-needed-in-sugh-region-for-tajik-children-/24693805.html (accessed 2014-09-12).

125

Mŭso Š. 2012-04-05. Nozir-i yakrav [in:] *Čarx-i Gardun* 2012-04-05.

Myakišev G.Ya., Buxovsev B.B. 2000. *Fizika. [sinf-i] 11.* Dušambe.

Nabi D. (date unknown). Čorrohao ba noziron-i BDA niyoz dorand [at:] *Jumhuriyat.tj* (website) http://jumhuriyat.tj/index.php?art_id=3965 (accessed 2011-03-28, cited as Nabi acc. 2011-03-28).

Nekrŭzov A. [2010]-10-13. Kinostudiya-i «Tožikfilm» somona-i interneti-i xud-ro boz kard [at:] *Tojnews.org* (website) http://tojnews.org/taj/index.php?option=com_content&task =view&id=9335&Itemid=42 (accessed 2011-10-06). 'Muslimah') [at:] http://nisaun.forumei.com/t120-topic (accessed 2011-03-25).

Nigori E., Munavvar D. (date unknown). Hameša imkon-i intixob hast [at:] *Jumhuriyat.tj* (website) http://jumhuriyat.tj/index.php?art_id=778 (accessed 2011-06-08).

Nisaun (forum). 2009-03-14 1:39 pm. Maslihatho baro-i ruzgor (*a post sent by 'Muslimah'*) [at:] http://nisaun.forumei.com/t197-topic (accessed 2011-03-25).

NOKIA. 2010-07-19. Dastur-i istifodabaranda NOKIA 6303I Classic. Nokia Corporation.

Normurod F., Qodiri S. 2005. *Fizika 7. Kitob-i darsi baro-i sinf-i 7.* Dušambe. Žam?iyat-i sahhomi-i nav?-i kušoda-i Matbu?ot.

Nuraliyon F. 2011-06-08. Vožanoma-i royonai [at:] http://www.parsi- tajiki.com/index. php?topic=3.0 (accessed 2011-07-23).

Nurob P. (date unknown). Zabon-i ilmi či muškilot dorad? [at:] Jumhuriyat (website) http:// www.jumhuriyat.tj/index.php?art_id=7305 (accessed 2014-09-12).

Oftob (website). (date unknown). Doxilkuni az fayl va xorižkuni ba fayl [at:] http://oftob.co m/%D0%BC%D0%B0%D0%B2%D0%BE%D0%B4%D0%B8-%D1%82%D0%BE% D2%B7%D0%B8%D0%BA%D3%A3/cpp-%D0%B4%D0%B0%D1%80%D1%81% D2%B3%D0%BE/35-cpp-file-io (accessed 2014-09-11).

Orifi Z. 2012-09-07. Beeline dar šabaka-i ižtimoi-i «odnoklassniki.ru» [at:] *Tojnews* (website) http://tojnews.org/node/3977 (accessed 2012-12-30).

Oriyonbonk (website). 2009. Bankomatho va POS-terminalho [at:] http://www.orienbank. com/index.php?t=5&i=165&l=2 (accessed 2011-03-23).

Özdemir D. 2002. Dezenfektanlara Direnç [at:] *Düzce Tıp Fakültesi Dergisi*, 2002; 4 (3). pp. 39-44.

Ozodagon (website). 2011-05-05. Ḡoratgari dar nazd-i mašžid… [at:] http://www.ozodagon. tj/khabarho/tojikiston/284-2011-05-05-1458.html (accessed 2011-10-08).

Ozodagon (website). 2011-06-09. Eron: Hadaf siloh-i hastai nest [at:] http://www.ozodagon. com/khabarho/jahon/789-2011-06-09-12-36-56.html (accessed 2013-01-16).

Ozodagon (website). 2011-07-20. Mullo Umar-ro az tariq-i SMS-ho «kuštand» [at:] http:// ozodagon.com/khabarho/jahon/1418-----sms--lr-.html (accessed 2011-08-08).

Ozodagon (website). 2011-07-22. 2 kilo-vu 922 gramm mavod-i muxaddir musodira gardid [at:] http://ozodagon.com/khabarho/tojikiston/1453.html (accessed 2011-08-10).

Parvina T. (date unknown). Nigohubin-i mŭyho dar fasl-i zamiston [at:] *Bonuvon.tj* (website) http://bonuvon.tj/index.php? option=com_content&view=article&id=215:nigohibi ni-muyho-dar- zimiston&catid=6:2010-01-19-12-48-47&Itemid=8&lang=ru&layout=d efault&month=2&year=2011 (accessed 2011-08-01).

Prezident-i Tožikiston (website). 2010-06-26. Xabarho [at:] http://www.president.tj/ habarho_260610b.html (accessed 2011-05-15).

Prezident-i Tožikiston (website). [2005]. Siyosat-i Xoriži, 15-18 noyabr-i sol-i 2005. Сиёсати хоричӣ, 15-18 ноябри соли 2005 [at:] http://www.president.tj/tunis.htm (accessed 2011-03-27).

Qadamova N. 2010-04-19. Telefon-i mobil saraton-i maḡzi sar mekunad [at:] http://www.millat.tj/ijtimo.html?start=330 (accessed 2012-11-11).

Qarḡizova A. 2009-07-07. Širkat-i «Somon Eyr» parvoz-i tayyoraho ba xatsayrho-i havoi-i nav-ro ba roh memonad [at:] *Asia-Plus* http://www.asiaplus.tj/tj/articles/54/3830.html# (accessed 2011-01-01).

Qayumzod A. 2008-10-10. Oḡoz-i muhokima-i dodgohi-i šikoyat-i «Somoniyon» [at:] *Radyo-i Ozodi* http://www.ozodi.tj/content/article/1328713.html (accessed 2011-05-15).

Qodir X. 2007-11-29. Barnoma-i hastai va hadaf-i noravšan-i Eron [at:] *Radyo-i Ozodi* http://www.ozodi.tj/content/article/752610.html (accessed 2011-07-10).

Quqanšoh A. 2011-01-19. Kortho-i plastiki va nafaqagiron-i nigaron [at:] *Millat* http://www.millat.tj/ijtimo/2253-korthoi-plastiki-va-nafaqagironi-nidaron.html (accessed 2011-01-31).

Radyo-i Ozodi. 2004-04-13. Naxustin guruh-i munaǯǯimon Donišgoh-i Milli-ro xatm mekunand [at:] http://www.ozodi.mobi/a/600228.html (accessed 2014-09-11).

Radyo-i Ozodi. 2005-05-11. Mole hast bo tamḡa-i "Made in Tajikistan"? at: http://www.ozodi.org/content/article/602347.html (accessed 2012-08-09).

Radyo-i Ozodi. 2005-11-30. Rangho-i nav-i oina-i nilgun [at:] http://www.ozodi.org/content/article/603849.html (accessed 2011-10-30).

Radyo-i Ozodi. 2006-04-06. Panǯnoma-i behtarin-i taronaho-i toǯiki (*recording*) [at:] http://audioarchive.rferl.org/ch9/2006/04/03/20060403-153000-TA-program.rm (accessed 2013-02-28).

Radyo-i Ozodi. 2006-04-25. Marz-i Toǯikiston-u Afḡoniston bo nurho-i lazeri didboni xohad šud [at:] http://www.ozodi.tj/archive/news/20060425/538/538.html?id=605187 (accessed 2011-03-25).

Radyo-i Ozodi. 2006-06-01. Xost-i Sozmon-i Milal baro-i muboriza bo SPID yo EYDZ. [at:] http://www.ozodi.org/content/news/605573.html (accessed 2011-06-11).

Radyo-i Ozodi. 2006-08-23. Dar suqut-i havopaymo 'se šahrvand-i Toǯikiston ǯon dodand' [at:] http://www.ozodi.org/content/news/606173.html (accessed 2011-09-20).

Radyo-i Ozodi. 2007-05-15. Oyo metavon dar Toǯikiston mavod-i radioaktivi-ro duzdid? [at:] http://www.ozodi.mobi/a/609046.html (accessed 2014-09-13).

Radyo-i Ozodi. 2007-11-21. Se namoišgoh-i baynulmilali dar Dušambe [at:] http://www.ozodi.org/content/article/719875.html (accessed 2011-11-26).

Radyo-i Ozodi. 2008-11-12. Siroyat-i angal-i SPID ba dahho kŭdak dar Uzbakiston [at:] http://www.ozodi.org/content/news/1348108.html (accessed: 2011-06-12).

Radyo-i Ozodi. 2009-02-13. Barxŭrd-i du mohvora dar kayhon [at:] http://www.ozodi.mobi/a/1492176.html (accessed 2014-09-11).

Radyo-i Ozodi. 2009-03-09. Tašannuǯ-i vaz? dar nimǯazira-i Koreya [at:] http://www.ozodi.org/content/article/1506730.html (accessed 2011-03-27).

Radyo-i Ozodi. 2009-03-10. Man?-i telefon-i hamroh dar ta?limgohho [at:] http://www.ozodi.org/content/article/1507219.html (accessed 2011-02-10).

Radyo-i Ozodi. 2009-04-20. Havopaymorabo ba polis taslim šud [at:] http://www.ozodi.org/Content/News/1612282.html (accessed 2011-01-06).

Radyo-i Ozodi. 2009-05-06. Toǯikiston čaroḡho-i nav-ro rŭšan mekunad [at:] http://www.ozodi.tj/content/transcript/1622722.html (accessed 2010-07-11).

Radyo-i Ozodi. 2009-10-07. Olimon-i IMA va Isroil sohib-i ǯoiza-i Nobel šudand [at:] http://www.ozodi.org/content/article/1845867.htm (accessed 2011-03-23).

Radyo-i Ozodi. 2010-01-25. Yak rŭz dar zer-i hiǯob [at:] http://www.ozodi.org/content/article/1939217.html?page=4&s=1&x=1#relatedInfoContainer (accessed 2011-03-24).

Radyo-i Ozodi. 2010-02-05. Radyo-i Ozodi dar mohvora [at:] http://www.ozodi.org/video/5541.html (accessed 2010-07-29).

Radyo-i Ozodi. 2010-07-26. Rahimzoda: Somoni-i qalbaki našr šudaast [at:] http://www.ozodi.tj/content/article/2109811.html?page=1 (accessed 2011-03-23).

Radyo-i Ozodi. 2010-07-15. Mažalla-i Bomdodi (broadcast).

Radyo-i Ozodi. 2010-07-21. Mažalla-i Bomdodi (broadcast).

Radyo-i Ozodi. 2010-07-27. Afzoiš-i muštariyon-i telefon-i mobili [at:] http://www.ozodi.org/content/article/2110629.html (accessed 2012-11-11).

Radyo-i Ozodi. 2010-08-13. Nerŭgoh-i atomi-i Bušahr ba kor oġoz mekunad [at:] http://www.ozodi.org/content/article/2126814.html (accessed 2011-07-09).

Radyo-i Ozodi. 2010-10-18. Faronsa az furŭš-i kištiho-i harbi ba Rusiya himoyat kard [at:] http://www.ozodi.org/content/news/2161339.html (accessed 2011-03-27).

Radyo-i Ozodi. 2010-11-10. Žasad-i oxirin qurboni-i suqut-i čarxbol paydo šud [at:] http://www.ozodi.org/content/article/2216213.html (accessed 2011-01-09).

Radyo-i Ozodi. 2011-03-08. Mažalla-i Bomdodi (broadcast).

Radyo-i Ozodi. 2011-06-23. Oġoz-i guftŭguho-i du raqib-i musallah bo bomb-i atomi. http://www.ozodi.org/content/news/24243663.html (accessed 2011-07-09).

Radyo-i Ozodi. 2011-12-01. Tožikiston: Nazarho dar bora-i iqdom-i sol-i navi-i širkatho-i mobili [at:] http://www.ozodi.org/audio/audio/335025.html (accessed 2013-01-03).

Radyo-i Ozodi. 2012-02-07. Rais-i "Tožiksodirotbonk" dar bora-i intiqol-i pul-i muhožiron (broadcast) [at:] http://www.ozodi.org/audio/audio/343145.html (accessed 2012-02-07).

Radyo-i Ozodi. 2012-07-17. Istifoda-i čarxbolho dar žangho-i Dimišq [at:] http://www.ozodi.org/archive/news/20120717/538/538.html (accessed 2013-01-31).

Radyo-i Ozodi. 2012-08-09. Sokinon-i Xoruġ: "Zindagi be telefon-i mobil muškil budaast" [at:] http://www.ozodi.org/content/article/24671622.html (accessed 2012-11-11).

Rahimov B., et al. 2006. *Fizika. Kitob-i darsi baro-i sinf-i 10.* Dušambe. Matbuot.

Raseloued (website). [2009-03-25]. furqat al-fidā? [at:] http://www.raseloued.net/Elfida.htm (accessed 2011-10-03).

Rasulov B. 2010-03-10. Organizmho-i az žihat-i genetiki takmildodašuda. ha yo na? [at:] [at:] *Millat* (website) http://millat.tj/ijtimo/1721-organizmhoi_az_chiha_995.html (accessed 2011-06-02).

Reporter (website). 2009-06-04. Gazetaho-i oyanda [at:] http://www.reporter.tj/index.php?option=com_content&view=article&id=282:2009-06-04-12-16-27&catid=20:reportertj (accessed 2010-07-29).

Referaty i kursovyje raboty besplatno (website). (date unknown). Muayyan kardan-i aktiviyat-i izotop [at:] http://refmegs.ru/22/dok.php?id=0011 (accessed 2011-07-30, cited as Referaty acc. 2011-07-30).

Rozi (Šaripov) T. 2011-10-19. Oyanda-i Tožikiston dar žomea-i ittilooti-i qarn-i asr-i XXI [at] Maktab (website) http://maktab.tj/news/oyandai-tojikiston-dar-jomeai-ittilootii-qarni-asri-xxi (accessed 2012-01-08).

Ruhulloh S. 2012-07-01 Barguzori-i Olimpiada-i čorum-i «NetRiders Tajikistan 2012» [at:] http://tojnews.org/node/3122 (accessed 2014-07-06).

Rŭzgor (website). 2011-03-01. Se bungoh-i xabari dar vodi-i rašt [at:] http://ruzgor.tj/matbuot/4547-se-bungohi-khabari-dar-vodii-rasht.html (accessed 2011-05-15).

Rŭzgor (website). 2011-04-27. Imrŭz niz az Dar?o sado-i tir-u tŭp šunida mešud [at:] http://ruzgor.tj/news/5037-imruz-niz-az-daryo-sadoi-tiru-tup-shunida-meshavad-.html (accessed 2011-05-29).

Růzgor (website). 2012-01-10. Aloqa-i bo-eʔtimod-i «megafon» dar růzho-i ǯašn [at:] http://ruzgor.tj/component/content/7014.html?task=view (accessed 2012-12-30).

Safar T. 2010-11-12. Ixtiro-i aǯib-i donišmandon-i toǯik [at:] http://www.ozodi.tj/content/article/2218511.html (accessed 2011-03-25).

Sahimov M.R. 2008. Bemoriho-i Invazioni-i hayvonot va parranda [at:] TVA (website) http://www.tva.tj/ru/index/index/pageId/113/ (accessed 2014-09-10).

Saidakbar L. 2012-01-19. Zimiston qotil-i xonabardůšon? [at:] Čarx-i Gardun (website) http://gazeta.tj/chg/4086-zimiston-1179otili-xonabard1263shon.html (accessed 2012-08-09).

Salmonov B. 2009-03-16. Dast-i tayyorasozi dar vatan ki bastaast? [at:] Asia-Plus (website) http://www.asiaplus.tj/tj/articles/276/3207.html# after: Minbar-i xalq №10, 2009-03-12 (accessed 2011-01-01).

Samsung. [2009]. *Samsung Warranty Card UWC0901*. Samsung Electronics.

Sang A. 2012-05-07. Ҳама чиз ва ҳама кас фаромӯш мешавад [at:] Faraj.tj (website) http://faraj.tj/life/561-1202ama-chiz-va-1203ama-kas-farom1263sh- meshavad.html (accessed 2014-09-10).

Sanginov F. 2010. Luḡat-i anglisi ba rusi-vu toǯiki va rusi ba anglisi-vu toǯiki. Grammatika. Dušambe 2010. Olam-i Kitob.

Satskaya P.N., Jamshedov P. 2000. English 10. (place unknown). Macmillan.

Satskaya P.N., et al. 2007. *Zabon-i Anglisi. Kitob-i darsi baro-i sinf-i 9*. Dušambe. Maorif va Farhang.

Saudi Food and Drug Authority (website). 2009-04-19. al-Brīyūnāt [at:] http://old.sfda.gov.sa/Ar/Food/Topics/food_quality_awareness/food-19052009-ar1.htm (accessed 2014-05-15, cited as SFDA 2009-04-19).

Serajtj.com (website). (date unknown). Raftor [at:] http://serajtj.com/raftor.htm (accessed 2011-03-24).

SFDA – *see* Saudi Food and Drug Authority

Shuhratjon. 2014-01-22. Taʔrix-i paydoiš-i kompyuter (MEX) [at:] toptjk (website) http://toptjk.com/reftj/informatics/print:page,1,437-tarihi-paydoishi-kompyuter-meh.html (accessed 2014-03-07).

Siddiqšoh A. 2014-06-09. "Bahoriston" – televizyon baro-i kůdakon yo…? [at:] Hizb-i Nahzat website http://www.nahzat.tj/2/item/12369-%E2%80%9C%D0%B1%D0%B0%D2%B3%D0%BE%D1%80%D0%B8%D1%81%D1%82%D0%BE%D0%BD%E2%80%9D-%D1%82%D0%B5%D0%BB%D0%B5%D0%B2%D0%B8%D0%B7%D0%B8%D0%BE%D0%BD-%D0%B1%D0%B0%D1%80%D0%BE%D0%B8-%D0%BA%D3%AF%D0%B4%D0%B0%D0%BA%D0%BE%D0%BD-%D1%91 (accessed 2014-07-05).

Sony. (date unknown). mulḥaqāt al-kāmīrā al-raqmiyya [at:] http://www.sony-mea.com/productcategory/accy-di-dsc-interface/view/retired?site=hp_ar_ME_i (accessed 2011-10-02).

Soros.tj (website). 2010. Dar Dušambe namoiš-i aksho tahti unvon-i «Toǯikiston ba sů-i ǯamʔiyat-i kušoda» oḡoz megardad [at:] http://soros.tj/index.php? option=com_content&view=article&id=59:2010-09-03-11-59-34&catid=32:2010-03-24-10-06-34 (accessed 2011-10-02).

Sufi Z. 2012-02-29. Agar bo durůḡsanǯ (detektor lži) bisanǯem (…) [at:] http://ozodagon.tj/8020-dur.html (accessed 2014-06-22).

Sulaymoni I. 2012-05-11. Firor-i maḡzho: muškil-e, ki dar Toǯikiston ǯiddi namegirand [at:] Ozodagon (website) http://www.ozodagon.com/2012/05/11/firori-mazo-mushkile-ki-dar-toikiston-idd-namegirand.html (accessed 2014-07-19).

Šarifov Ʒ. (date unknown). Farhang-i texniki ʒuzʔ-i huviyat-i millist [at:] *Jumhuriyat* (website) http://www.jumhuriyat.tj/index.php?art_id=5271 (accessed 2012-12-30).

Šarifov X. (date unknown). Yodnoma-i Mirzo Tursunzoda [at:] http://www.cit.tj/mirzot/yodnoma2.php?id_sher=39 (accessed 2014-09-11).

Šerali L. [2009]. Ġazalho-i Loiq Šerali – Bo ovoz-i asli-i šoir [at:] *Termcom* (website) http://termcom.tj/index.php?menu=bases&page=loiq_music&lang=toj (accessed 2014-07-20).

Šikebo S. 2010-08-12. Murod Kambaġali [at:] *Čarx-i Gardun* http://www.gazeta.tj/chg/209-murod-kamba1170all250.html (accessed 2011-03-25).

Šodmonov X. (date unknown). Oits [at:] doctor.uz (website) http://doctor.uz/articles/kasal liklar/oits.html (accessed 2011-06-12, cited as Šodmonov acc. 2011-06-12).

Tabnak (website). 1390-03-31 HŠ. doruq-sanʒi-ye Āber-bānk [at:] http://www.tabnak.ir/fa/news/172273/%D8%AF%D8%B1%D9%88%D8%BA%E2%80%8C%D8%B3%D9%86%D8%AC%DB%8C-%D8%B9%D8%A7%D8%A8%D8%B1%D8%A8%D8%A7%D9%86%DA%A9 (accessed 2011-10-06).

Tajik Air (website). (date unknown). Korxona-i Vohid-i Davlati-i Havopaymoi [at:] http://www.tajikair.tj/taj/index.php?option=com_content&task=blogcategory&id=14&Item id=107 (accessed 2010-12-30).

Tajik Air (website). (date unknown). Dar bora-i širkat-i havopaymoi [at:] http://www.tajikair.tj/taj/index.php?option=com_content&task=view&id=90&Itemid=61 (accessed 2011-03-27).

Tcell (website). (date unknown). BlackBerry® Curve TM 8520 [at:] http://tcell.tj/tg/index/index/pageId/805/ (accessed 2012-04-15).

Tcell (website). (date unknown). Bastaho-i Tarofaho [at:] http://tcell.tj/tg/index/index/pageId/2324/ (accessed 2014-09-11).

Tcell (website). 2011-07-18. Dar hamkori bo širkat (…) [at:] http://www.tcell.tj/tg/index/index/pageId/1211/ (accessed 2012-12-30).

Tebyan (website). 1392-03-09 HŠ. Āškārsāz-e Gāyger-Muler (Gieger-muller counter *sic*) čist? [at:] http://www.tebyan.net/newindex.aspx?pid=246434 (accessed 2014-07-25).

Termcom.tj (website). (date unknown). Qoidaho-i soziš [at:] http://www.termcom.tj/?menu=about&page=art_about_4&lang=toj (accessed 2010-07-28).

tib.islom.uz (website). 2011-04-23. Hinno – ažoyib vosita [at:] http://tib.islom.uz/islomiy-tib/90-maqolalar.html (accessed 2011-09-14).

Toʒikon (website). (date unknown). Salomati-i čašm hangom-i kor bo royona [at:] http://www.tojikon.org/tojiki/index.php?option=com_content&task=view&id=117&Itemid=23 (accessed 2010-10-06).

Toʒik Sodirot Bonk (TSB). (date unknown). Kortho-i plostik [at:] http://www.brt.tj/tj/corporate/cards/ (accessed 2014-06-22).

Tojnews (website). 2012-10-12. ISIS megūyad, Eron tavon-i bunyod-i siloh-i atomi-ro paydo kardaast [at:] http://www.tojnews.org/node/4468 (accessed 2014-08-20).

Tojnews (website). 2013-04-01. Ešoni Nuriddinʒon: Dar bora-i taloqho-i "sms"-i [at:] http://www.tojnews.org/eshoni-nuriddinjon-dar-borai-taloqhoi-smsi (accessed 2014-06-22).

Tojnews (website). 2013-07-08. Prezident ba yatimon televizor, yaxdon, qolin, čangkaš va planšet tuhfa kard [at:] http://tojnews.org/ishtiroki-president-dar-tuyi-khayriyai-oilahoi-kambizoat-dar-khatlon (accessed 2014-09-11).

Tojnews (website). 2013-07-12. Kumita-i televizyon va radyo muntazir-i huʒʒatho-i hnit [at:] http://tojnews.org/kumitai-tv-muntaziri-hnit (accessed 2013-12-10).

Turaǯonzoda E.M. 2007-07-05. Hiǯob az ogoh-i taʔrix to kunun [at:] http://www.turajon.org/articles_view.php?id=13 (accessed 2013-01-16).

UNESCO (website). (date unknown). al-faḍāʔ wa handasat al-milāḥa al-ǧawwiyya [at:] http://www.unesco.org/ar/earthsciences/themes/space-education/space-and-aeronautic-engineering/ (accessed 2011-09-24).

Usmoniyon Ǯ. 2012-04-30. Vazir-i oyanda-i energetika az Danḡara, ammo loiq [at:] Millat (website) http://millat.tj/component/content/3073.html?task=view (accessed 2014-09-16).

Ustoyev M.B. 2008. *Biologiya. Anatomiya va fizyologiya-i odam. Kitob-i darsi baro-i sinf-i 9.* Dušambe. Sobiriyon.

Vafobek O. (date unknown). Bemoriston-i bayninohiyavi-i bemoriho-i sil ba istifoda doda mešavad [at:] Jumhuriyat (website) http://www.jumhuriyat.tj/index.php? art_id=4435 (accessed 2014-09-07).

Vahdat11. 2010-10-26. Šomārešgar-e Gāyger-Muler. [at:] http://vahdat11.blogfa.com/post-125.aspx (accessed 2011-07-30).

Vazorat-i Korho-i Doxili-i Ǯumhuri-i Toǯikiston (website). 2013-04-22. Natiǯa-i amaliyot-i "Čaroḡak-i rahnamo va yak moh-i bexatari" [at:] http://www.mvd.tj/index.php/tj/asosi/1680-nati-ai-amalijoti-charo-aki-ra-namo-va-yak-mo-i-bekhatar (accessed 2014-02-02, cited as VKD 2013-04-22).

Vazorat-i Korho-i Doxili-i Ǯumhuri-i Toǯikiston (website). 2014-01-08. Duzdi-i kompyuter-ho az maktab soxta šud [at:] http://www.vkd.tj/index.php/tj/asosi/3317-duzdii- kompyuter-o-az-maktab (accessed 2014-09-11, cited as VKD 2014-01-08).

Vazorat-i Korho-i Doxili-i Ǯumhuri-i Toǯikiston (website). (date unknown). *Home page* [at:] http://www.mvd.tj (accessed 2011-10-06, cited as VKD acc. 2011-10-06).

Vazorat-i Naqliyot va Kommunikatsiya ǮT. (2009-03-30). Maǯlis-i nazorati (…) [at:] http://www.mintranscom.tj/index.php?option=com_content&task=view&id=102&Itemid=112 (accessed 2010-01-01, cited as VNK 2010-01-01).

Viloyat-i Suḡd – *see* Maqomot-i Iǯroiya-i Hokimiyat-i Davlati-i Viloyat-i Suḡd.

VKD – *see* Vazorat-i Korho-i Doxili-i Ǯumhuri-i Toǯikiston.

VNK – *see Vazorat-i Naqliyot va Kommunikatsiya ǮT.*

Voy. 2009-12-29. Yevrokomissiya ovozni cheklamoqchi [at:] http://www.voy.uz/news/yevrokomissiya_ovozni_cheklamoqchi/2009-12-29-3852 (accessed 2011-02-20).

Warior[76] 2012-11-02. Mušak, Klaviatura va Printer [at:] http://omuzgor.my1.ru/publ/referat-kho_bo_zaboni_tochiki/asos_oi_tekhnikai_kompjuter/ mushak_va_klaviatura/6-1-0-28 (accessed 2013-02-17).

Wikipedia (website). (date unknown). Mobil telefon [at:] http://uz.wikipedia.org/wiki/Mobil_telefon (accessed 2011-02-11).

Wikipedia (website). (date unknown). Nīldāṅt [at:] http://ur.wikipedia.org/wiki/%D9%86%DB%8C%D9%84%D8%A7%D8%AF%D8%A7%D9%86%D8%AA (accessed 2011-10-09).

Wikipedia (website). (date unknown). Ǯahān-gostār [at:] http://fa.wikipedia.org/wiki/%D9%88%D8%A8_%D8%AC%D9%87%D8%A7%D9%86%E2%80%8C%DA%AF%D8%B3%D8%AA%D8%B1 (accessed 2011-10-06).

Wikipedia (website). (date unknown). Veb-bin [at:] http://fa.wikipedia.org/wiki/ %D9%88%D8%A8%E2%80%8C%D8%A8%DB%8C%D9%86 (accessed 2011-01-25).

---

[76] Sic!

Xalili I. 2011-07-14. Bongi xatar: Insulin nest! [at:] Ozodagon (website) http://www.ozo dagon.com/7708-bongi-hatar-insulin-nest.html (accessed 2014-09-12).

Xayrov X.S. (date unknown). Xůrok-i parxezi baro-i bemoron-i diabet-i qandi novobasta az insulin [at:] Vazorati tandurusti va hifz-i ižtimoi-i aholi-i ЗТ (website) http://www.health. tj/kh%D3%AFroki-par%D2%B3ez%D3%A3-baroi-bemoroni-diabeti-%D2%9Bandi-novobasta-az-insulin (accessed 2014-07-30).

Yak Darveš (blog). 2008-05-20. Kindle Up Our Hopes [at:] http://dariussthoughtland. blogspot.com/2008_05_01_archive.html (accessed 2011-07-23).

Yunusi M. 2007. *Elementho-i informatika baro-i hama*. Dušambe. Donišgoh-i Davlati-i Mil-li-i Tožikiston.

Zaman Türkmenistan. 2009-06-16. Meşhur telefonlaryň onusy [at:] http://www.zamantm. com/tm/newsDetail_getNewsById.action;jsessionid=2340DAA6604155F05912586C1 7A449E3.node1?newsId=3076 (accessed 2011-02-20).

Zare'i V. 1389-09-10 HŠ. Veb-e žahān-gostar World Wide Web [at:] http://irmeta.com/meta/ b622/t5384/ (accessed 2014-09-27).

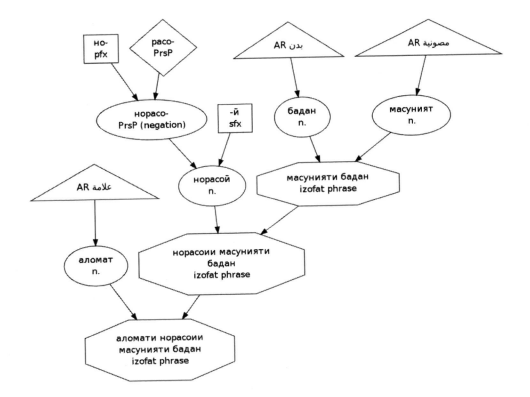

Fig. 1

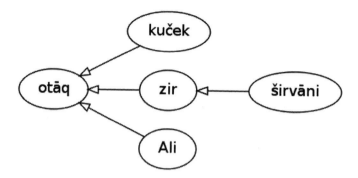

Fig. 2

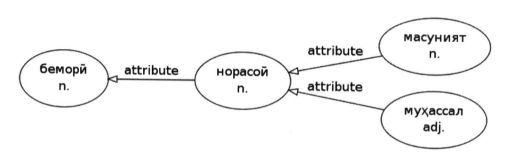

Fig. 3

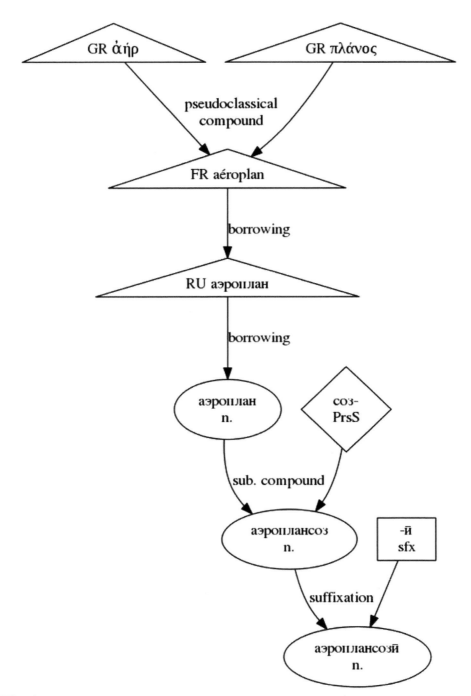

Fig. 4

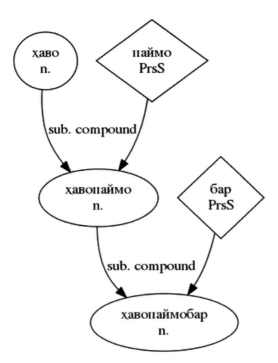

Fig. 5

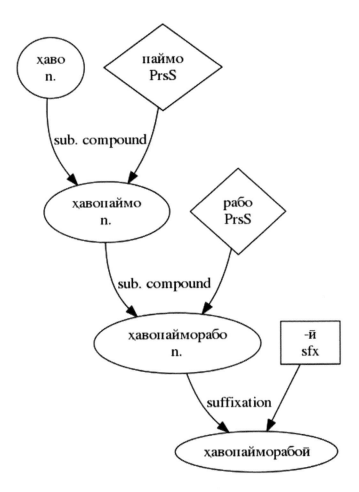

Fig. 6

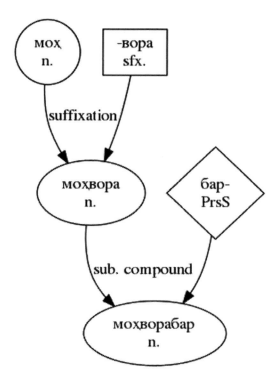

Fig. 7

Fig. 8

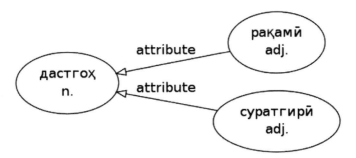

Fig. 9

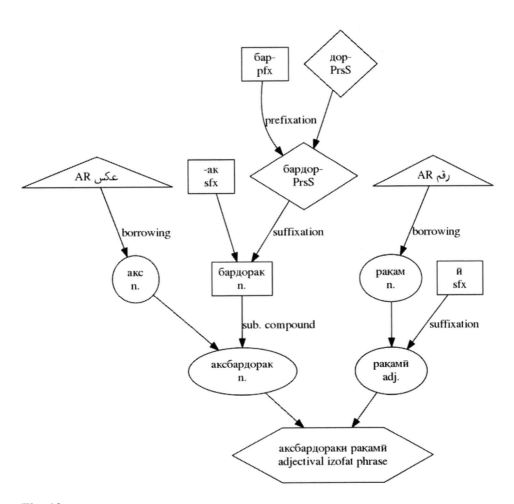

Fig. 10

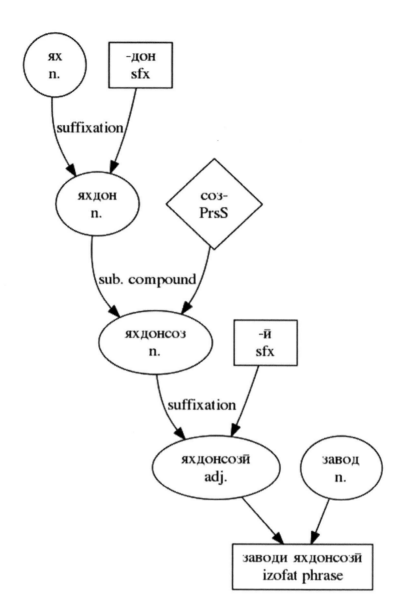

Fig. 11

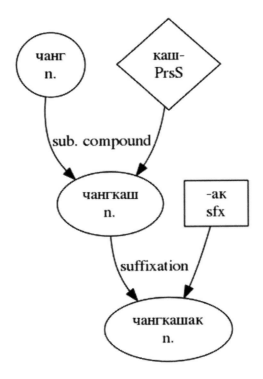

Fig. 12

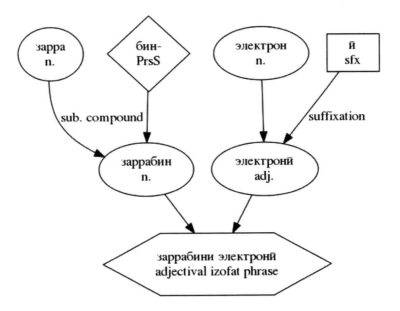

Fig. 13

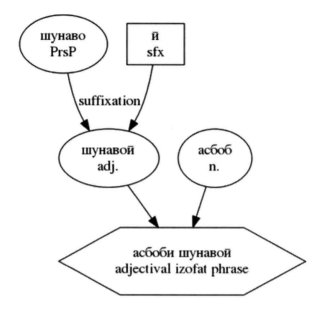

Fig. 14

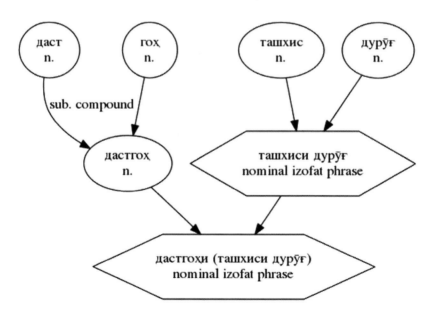

Fig. 15

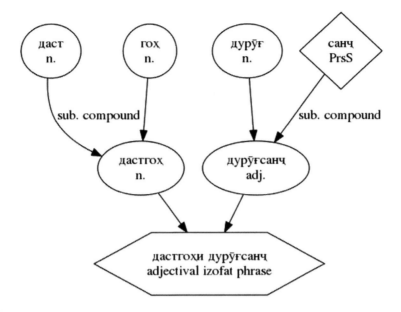

Fig. 16

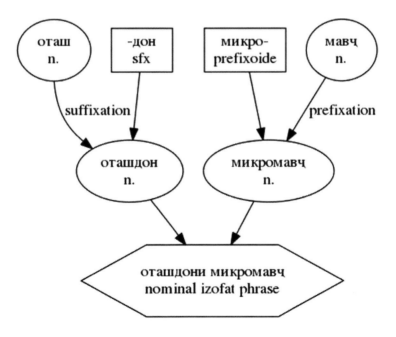

Fig. 17

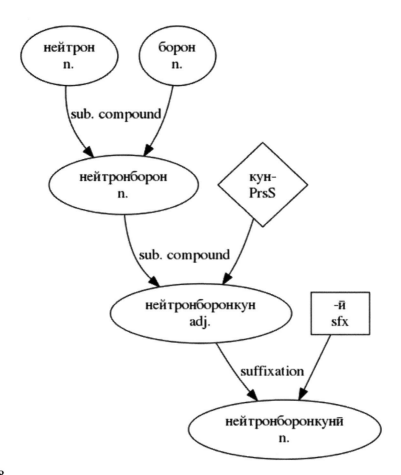

Fig. 18

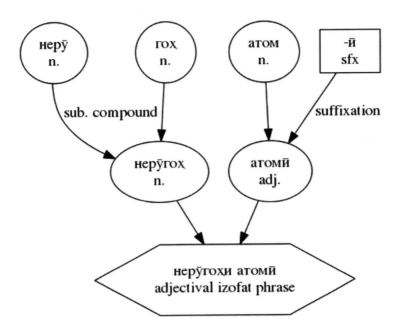

Fig. 19